# Hazardous Gases and Fumes

## Peter Warren BSc., Ph.D.

Health and Safety Consultant
Eastman Dental Institute and Hospital
London WC1X 8LD

BUTTERWORTH
HEINEMANN

Butterworth-Heinemann
Linacre House, Jordan Hill, Oxford OX2 8DP
A division of Reed Educational and Professional Publishing Ltd

℞ A member of the Reed Elsevier plc group

OXFORD   BOSTON   JOHANNESBURG
MELBOURNE   NEW DELHI   SINGAPORE

First published 1997

**British Library Cataloguing in Publication Data**
Warren, Peter, 1939–
    Hazardous gases and fumes
    1. Hazardous substances      2. Hazardous substances – Health
    aspects
    I. Title
    363.1'7

ISBN 0 7506 2090 0

**Library of Congress Cataloguing in Publication Data**
Warren, Peter J.
    Hazardous gases and fumes/Peter J. Warren.
    p.      cm.
    Includes index.
    ISBN 0 7506 2090 0
    1. Hazardous substances.      2. Gases, Asphyxiating and poisonous.
    3. Industrial safety.      I. Title.
    T55.3.H3W37                                              96–50401
    604.7–dc21                                                   CIP

Composition by Genesis Typesetting, Rochester, UK
Printed and bound in Great Britain by
Biddles Ltd, Guildford and King's Lynn

# Contents

# Preface

The purpose of this book is to provide the reader with a concise summary of the physical properties, detection methods, handling precautions, and the biochemical and physiological effects of gases and fumes found at work, so that they may be handled safely. These substances vary considerably in their chemical, physical and physiological properties.

Some gases such as oxygen are essential to life, since the oxidation of many substances in biological systems ultimately requires the presence of molecular oxygen to form water, under the influence of enzyme catalysts such as the oxidases.

Other gases such as nitrogen do not take an active role in biological systems, but they are important in industrial processes, e.g. the manufacture of ammonia.

A third group might be termed the toxic gases or fumes. These are often important volatile chemicals used in industry, but they possess properties that demand very careful handling and safety precautions during their use particularly in industrial processes.

Defining what is meant by a gas is not easy. At the simplest level it is a chemical substance that is volatile at a stated range of temperatures and has no definite shape or volume of its own.

Gases can be produced as a result of heating a solid, evaporation from a liquid, as a result of chemical reactions, or directly from a surface of a metal either at room temperature, as in the case of mercury, or at very high temperatures in the form of gaseous fumes.

Gases and fumes are a very diverse series of substances. Many are simple molecular forms of elements such as hydrogen $H_2$ or nitrogen $N_2$. Other elemental forms are extremely reactive particularly at high temperatures, e.g. chlorine $Cl_2$ or fluorine $F_2$. Gases that exist as compounds of elements can have varying properties, e.g. nitrogen oxide gases existing as nitrous oxide $N_2O$, nitric oxide NO, nitrogen dioxide $NO_2$ or $N_2O_4$ depending on the temperature.

Fumes can arise from a variety of chemical and industrial processes. The fabrication of metal using high temperature gas or arc welding produces toxic gases from the welding process, and metal or metal oxide fumes. The effects of exposure to fumes may be immediate, e.g. cyanide, or delayed, e.g. in the case of many metal fumes leading to non-specific fever.

Gases have very different chemical, physical and toxicological properties for humans. Nitrous oxide is used extensively as an anaesthetic gas in surgery, and nitrogen dioxide is a corrosive and fatally toxic gas. This was demonstrated some years ago when a consignment of gas cylinders of nitrous oxide was found to contain a significant quantity of toxic nitrogen dioxide, which led to the death of a patient, who unfortunately received this gas mixture.

This book attempts to give a clear and comprehensible presentation of the data required to use and handle gases and fumes safely, to give practical information on the methods of detection of the substances to enable exposure limits to be monitored, and to offer guidance on the safety precautions and occupational health considerations required in the workplace.

For each gas and fume described, basic first aid treatment is described for persons affected by inhalation, ingestion, eye or skin contact with the toxic substance.

# Acknowledgements

I would like to thank British Rail for providing the opportunity for me to draft the outline of this book on my train journeys to Manchester. Its origins lay in many discussions I have had with academic and technical experts in the fields of occupational medicine, toxicology and the fire services. In particular, I am indebted to the long discussions and helpful advice from the late Dr Donald Hunter CBE, MD, DSc., FRCP on occupational health and safety, and to the colourful and enthusiastic fireman, Deputy Acting Chief Officer Charles Clisby MBE, QFSM of the London Fire Brigade, who was one of the originators of the now widely used Hazchem Coding for toxic materials, pioneered by the Greater London Council.

Without the help of my son James Moore-Warren, I would have been unable to master the difficulties of the word processor and the presentation of this work.

I am grateful to all the staff of Butterworth-Heinemann for their patience and help in producing the book.

# Abbreviations used in the text

## Units of measurement

| | |
|---|---|
| atmos. | the pressure of a gas measured in atmospheres. 1 atmosphere = $101\,325\,\text{Nm}^{-2}$ which is approximately $10^5\,\text{Nm}^{-2}$. |
| kg | kilograms, 1000 grams. |
| $\text{mg/m}^3$ | milligrams of a substance per cubic metre of air or vapour. |
| mmHg | millimetres of mercury (used clinically as units of pressure). |
| N | unit of force known as the newton. It is the force required to produce an acceleration of 1 metre per second per second in a mass of 1 kilogram. |
| ng | nanograms, $10^{-9}$ grams. |
| nm | nanometres, $10^{-9}$ metres, used to express wavelengths of light. |
| Pa | unit of pressure, known as the pascal, which is the force 1 newton exerts over a surface of 1 square metre. |
| ppb | parts of a substance per billion (= thousand million) parts of air or liquid. |
| ppm | parts of the substance per million parts of air or a liquid. |
| µg | micrograms, $10^{-6}$ grams. |
| um | micrometres, $10^{-6}$ metres, used to express wavelengths of light. |
| Ci | unit of radioactivity (the Curie) = $3.7 \times 10^{10}$ nuclear transformations per second (now superseded by the Becquerel (Bq) = 1 nuclear transformation per second). |

## Physicochemical properties

| | |
|---|---|
| D | Density measured in $\text{kg/m}^3$. |
| EL | Explosion limits (upper and lower) as % by volume of the substance in air. |
| FP | Freezing point in degrees Celsius. |
| IT | Autoignition temperature expressed as % by volume of the substance in air. |
| LEL | Lower explosion limit expressed as % by volume of the substance in air. |
| MP | Melting point in degrees Celsius. |
| MW | Molecular weight of mass. |
| SG | Specific gravity measured in $\text{kg/m}^3$. |
| VD | Vapour density in mmHg and Pa units. |
| VP | Vapour pressure in mmHg and Pa units. |

## Exposure standards

COSHH      Control of Substances Hazardous to Health (COSHH) Regulations 1994.

OES      Occupational Exposure Standard. This is the concentration of an airborne substance averaging over a reference period, usually 40 hour working week, at which there is no evidence of injury to the exposed person. If a substance is assigned an OES value, exposure by inhalation should be reduced to or below that value.

STEL      Short-Term Exposure Limit, i.e. the maximum concentration to which a worker may be continuously exposed for a period of up to 15 minutes, provided that:
(a) no more than four STEL periods are permitted in each working day,
(b) 1 hour elapses between each STEL period,
(c) the daily TWA is not exceeded.

TWA      Time Weighted Average concentration of an airborne substance for a normal 8 hour working day (40 hour working week).

TWA(C) or (C)      This value represents a **ceiling concentration which must not be exceeded**.

## Safety organizations

ACGIH      American Conference of Governmental Industrial Hygienists.

EOHS      Encyclopaedia of Occupational Health and Safety published by the ILO.

HSE      Health and Safety Executive of the United Kingdom Health and Safety Commission.

IAEA      International Atomic Energy Agency.

ICRP      International Commission for Radiation Protection.

ILO      International Labour Office (Geneva), a part of the United Nations World Health Organization.

OSHA      Occupational Safety and Health Administration of the United States Government.

## Miscellaneous

AFFF      Aqueous film forming foam.

CABA      Compressed air breathing apparatus.

EDTA      Ethylenediamine tetraacetic acid.

# Glossary of medical terms

The following medical terms listed below have been used in the Occupational Health sections of this book. These simplified definitions may be of help to readers who are not familiar with them. A fuller definition can be found in *Butterworths Medical Dictionary*, 2nd Edition, Macdonald Critchley (ed.), Butterworths (1978).

**Allergic reactions**
Reactions produced as a result of the action of an allergic inducing agent on the body.

**Analgesia**
The process by which the sensation of pain is reduced.

**Angiosarcoma**
A cancerous tumour affecting the blood or lymphatic vessels.

**Anoxia**
This literally means without oxygen. The term is often used to describe hypoxia which is a state of inadequate oxygenation of the tissues.

**Anuria**
The complete cessation in the production of urine by the kidneys.

**Apnoea**
This is cessation in breathing. Apnoea may follow a period of hyperventilation by the subject causing an excessive removal of carbon dioxide from the blood by the lungs. Normal breathing is restored when the normal carbon dioxide level is established.

**Asthma**
A general term used for a syndrome characterized by paroxysmal attacks of coughing and shortness of breath caused by the narrowing of the small bronchi and bronchioles in the lung, the swelling of the mucous membrane and exudation of mucus.

**Ataxia**
The inability of the subject to coordinate normal walking movements due to an effect on the nervous system. Ataxic gait is seen when there is damage to the cerebellum or basal ganglia in the brain.

**Auronasal mask**
This term is used to describe a face mask or respirator which protects only the nose and the mouth of the subject.

**Autoimmune response**
These are events related to an immune reaction developed in response to antigens from the subjects own tissues.

**Basal ganglia**
A group of nerve cells situated in the mid brain and responsible for the relaying of nerve impulses from the brain cortex to the motor tracts of the spinal cord. Damage to the basal ganglia is usually seen as interference in normal muscle movements and coordination.

**Bronchiolitis**
Inflammation of the bronchioles in the lung.

**Bronchitis**
Inflammation of the mucous membrane of the larger and medium-sized bronchi. Bronchitis can be caused by many agents particularly toxic and irritant gases and fumes.

**Carcinomas (endothelial type)**
A cancer developing in the lining cells of the body cavities and/or the blood vascular system.

**Carcinomas (mesenchymal type)**
A cancer developing in cells of the mesoderm which gives rise to bone, cartilage, connective tissue as well as the blood and lymphatic system.

**Cardiac fibrillation**
A disorder in cardiac rhythm in which the atria or the ventricles of the heart produce spontaneous uncoordinated contractions.

**Chelating agents**
A chemical molecule that can readily combine with a toxic metal ion to produce a stable chelated product that may be excreted from the body. Calcium disodium versenate readily combines with $Pb^{2+}$ ions in the blood to form a lead versenate complex which is readily excreted in the urine.

**Chemical pneumonitis**
Inflammation of the lung caused by the inhalation of an irritant chemical substance.

**Conjunctivitis**
The irritation of the conjunctiva of the eye leading to inflammation and swelling of the eye and surrounding tissues.

**Corneal damage**
Damage to the corneal covering of the eye lens produced by physical or chemical injury.

**Cyanosis**
The physical appearance of the subject that is deficient in oxygenated blood exhibiting blue lips, nail beds and conjunctivae.

**Dermatitis**
Inflammation of the skin caused by irritation from chemical or biological agents, ultraviolet, X-ray or radionuclide radiation or from hypersensitization caused by an allergic skin response.

**Diuresis**
The production of excessive quantities of urine by the kidneys leading to the passage of large volumes of urine from the bladder.

**Duodenum**
Part of the gastrointestinal system connecting the stomach with the upper part of the ileum or small bowel.

**Dyspnoea**
A state of raised lung ventilation rate usually as a result of inadequate uptake of oxygen from the air in the lung alveoli into the pulmonary blood supply. Poor oxygen uptake by the lungs can be as a result of low oxygen concentration in the inspired air, inflammation and/or oedema in the lungs.

**Eczematoid dermatitis**

An eczema like scaling or oozing inflamed eruption of the skin produced by an infection or chemical agent.

**Emphysema**

A chronic clinical condition causing the dilation of the alveoli which are the small terminating sacks at the end of the bronchiole tree in the lungs.

**Epithelia**

The cells lining organs in the body, e.g. lung epithelia or gastrointestinal epithelia.

**Exchange transfusion**

The process by which the blood from the affected person containing toxic substances can be simultaneously exchanged with an equal volume of normal blood containing compatible blood groups.

**Extra pyramidal system**

The system of motor or efferent nerves lying outside the pyramidal tract of the spinal cord, e.g. motor nerves supplying the skeletal muscles.

**Free radical**

An uncombined highly reactive chemical radical, often formed in a vigorous oxidative process, which can produce tissue damage if not combined with protecting polyunsaturated organic molecules such as vitamin E or A.

**Gastric lavage**

Washing out the stomach and removing its contents using a gastric tube introduced into the stomach. This procedure is used if a toxic chemical or poison is swallowed and helps to prevent the harmful substance from passing into the ileum where efficient absorption can take place.

**Gout**

A disease of purine metabolism characterized by attacks or arthritis associated with raised serum uric acid levels. Gout is found in chronic lead poisoning, which is also associated with renal failure.

**Haemodialysis**

The process by which the blood from the affected person containing toxic substances can be simultaneously exchanged with an equal volume of normal blood containing compatible blood groups.

**Haemoglobin**

A respiratory metalloprotein present inside the red cells of the blood, responsible for the transport of oxygen from the lungs to the tissues.

**Haemolysis**

The process of breakdown of the red cell membrane leading to the release of haemoglobin into the plasma which can lead to blockage of the filtration glomerular membrane in the kidney.

**Haemolytic agent**

A chemical or biological substance which causes the breakdown or rupture of the red cell membrane leading to the loss of the pigment haemoglobin.

**Haemoptysis**

The coughing up of significant quantities of blood in the sputum.

**Haemorrhage**

A loss of blood caused by a physical or chemical trauma to body tissues.

**Hyperreflexia**

A condition in which neuromuscular reflexes are greatly exaggerated.

**Methaemoglobin**

The reduced chemical form of haemoglobin

**Myocardial depression**

A depression of activity in the action of the heart muscles, usually those of the left ventricle.

### Narcosis
A stupor induced by anaesthetic gases or drugs tending to produce insensibility and/or paralysis.

### Nephrotoxic agent
Chemical or biological substances that can damage the tissues or functioning of the kidneys.

### Oedema
The presence of excessive amounts of fluid in the intercellular tissue spaces in the body.

### Olfactory fatigue
The condition in which an odour originally perceived by the nose cannot be detected by smell.

### Parathesia
A condition affecting the afferent peripheral nerves particularly in the hands and feet producing a sensation of pins and needles.

### Peritoneal dialysis
In this form of dialysis, isotonic salt solutions are perfused into the body's peritoneal cavity so that using the peritoneal wall as a dialysis membrane, toxic substances can be removed from the blood vessels in the peritoneal wall into the perfusing fluid, and then removed from the body via a drainage tube.

### Polyneuritis
This is inflammation of many nerves describing multiple and peripheral neuritis. The early clinical effects of neurotoxic agents are often seen as a generalized polyneuritis.

### Protein denaturing
A process by which the normal physical structure of a protein is altered irreversibly. Physical, chemical and biological agents can denature proteins changing their physical state and biological function.

### Pulmonary hyperreactivity
A condition occurring in the lungs where the lung tissue becomes sensitized to the presence of a chemical or biological agent, producing severe irritation of the mucosa leading to coughing or wheezing. Severe hyperreactivity can lead to an asthma type attack.

### Pulmonary oedema
The build-up of fluid in the lungs usually caused by infection of chemical irritation of the tissues lining the lung. Pulmonary oedema can give rise to dyspnoea, a form of respiratory distress due to a four-fold increase in the lung ventilation rate. This is caused by the inability to oxygenate blood flowing through the pulmonary circulation.

### Reflex coughing
The stimulation of the cough reflex in the lung following the introduction of a foreign particle, irritant gas or fume.

### Renal failure
The failure of the functioning of the kidneys to filter the blood and produce a urine of normal composition.

### Siderosis
Deposits of fine particles of iron in the lymphoid aggregations of the lungs occurring in miners, welders and other metal workers.

### Sub-acute combined degeneration of the spinal cord
A clinical condition in which the patient exhibits neurological symptoms associated with damage to the spinal cord. Damage to the ascending spinal tracts can produce loss of sensations such as touch, pain and balance. Damage to the descending spinal tracts can produce full or partial paralysis of the muscles.

**Tracheitis**
Inflammation of the membrane lining the trachea.
**Upper respiratory tract**
Anatomically, this is a region represented by the upper airways, namely the trachea and bronchi, which lead to the upper part of the respiratory tract consisting of small bronchioles and alveoli.
**Urinary alkalinization**
The process by which urine, produced by the kidneys, is rendered alkaline by orally administering salts such as citrates or tartrates which are metabolized in the body to give alkaline excretory products.
**Vasoconstriction**
Constriction of the blood vessels usually caused by the contraction of the circular muscles in the walls of the blood vessels. This process can be induced by neural or chemical factors.

# General references

The references listed below are useful to the technical sections in the book.

1. Eleanor Berman
   *Toxic metals and their analysis*, Heyden and Son Ltd, London (1980).
   Professor Berman's book deals with the metabolism, toxicology and methods of analysis of 31 important metals. ⌄

2. William Braker, Allen L. Mossman, and David Seigel
   *Effects of Exposure to Toxic Gases – First Aid and Medical Treatment*. Second Edition, Matheson, Lyndhurst, New Jersey (1977).
   This book specializes on the safety precautions and preventative measures needed when using toxic gases in industrial locations.

3. Peter B. Cook
   *Trevethick's Occupational Health Hazards*, Second Edition, Heinemann Medical Books (1989).
   Dr Cook's monograph deals with chemical and physical hazards found in the workplace, including toxic gases and fumes.

4. BOC Gases
   *Safety Data Sheets for Gases*, produced by BOC Gases, Priestley Road, Worsley, Manchester M28 2UT.
   These Safety Data Sheets are produced by this company for all gases supplied in cylinders or as cryogenic liquids. They are extremely useful as a rapid guide to the hazards and properties of various gases.

5. Health and Safety Executive
   *EH40/95 Occupational Exposure Limits*. HSE Books, Suffolk, England (1995).
   This short manual is prepared by the HSE and contains lists of Occupational Exposure Standards, and maximum exposure limits as part of the requirements for the Control of Substances Hazardous to Health Regulations 1994.

6. Donald Hunter
   *The Diseases of Occupations*, The English Universities Press Ltd, London (1975).
   Dr Hunter's book, written by a distinguished occupational physician, probably the father of British occupational medicine, describes the industrial processes and causes that have led to occupational-induced diseases, much of the work from personal observations.

7. Kurt Leichnitz
   *Detector Tube Handbook – Air investigations and technical gas analysis with Drager tubes*, Seventh Edition, Dragerwerk Aktiengesellschaft, Lubeck, Republic of Germany (July 1989).
   In the form of a pocket book, this edition gives full information on the wide range of chemical reaction detector tubes produced by Drager.

8. Henry Matthew and Alexander A.H. Lawson
   *Treatment of Common Acute Poisoning*, Fourth Edition, Churchill Livingstone, Edinburgh, London, and New York (1979).
   This book is intended for accident and emergency units that have to deal with cases of acute poisoning. Chapter 16 deals with acute metal poisoning.
9. *Encyclopaedia of Occupational Health and Safety*, Third Revised Edition, Dr Luigi Parmeggiani (ed.), International Labour Organization, International Labour Office, Geneva, Switzerland (1991).
   This two volume encyclopaedia represents an extremely valuable and comprehensive account of the hazards and safety requirements for the handling of all types of toxic materials. The occupational health problems and control measures to minimize the effects of exposure are fully covered.

# Part I
# Methods of measuring gases with gas detection equipment

Chemical and physical methods are used to determine the type and concentration of gases present in the air and solutions. Analysis of the environmental air is normally considered to be the most appropriate way of determining the degree of possible exposure to a gas. This method is preferred to analysis of the gas present in the blood or appropriate excretory products formed within the body and present in the urine or faeces, as it gives data on possible exposure of other workers and/or members of the public.

The various methods available for the analysis of gases can be summarized as follows:

## 1. Atomic absorption spectroscopy

Gases and volatile liquids can be introduced into acetylene/air or propane/air flames which produce a mixture of atoms and ion species of the particular gas. The atomic form of the gas absorbs strongly at a particular wavelength specific to the element present. If a beam of light at this wavelength is passed through the flame, via a monochromator to a photomultiplier detector, the electrical output from the detector can be amplified and displayed by a chart recorder. The value of the atomic absorbance measured is proportional to the concentration of the gas passing through the flame. This method can detect atom species in mixture at the parts per million level.

Recently, non-flame and specially constructed carbon tube furnaces have replaced the traditional flame system and have led to the development of methods giving high sensitivity, greater accuracy, and with automated sample analysis.

## 2. Balmer emission spectral analysis

In this method the gas or vapour sample to be analysed is placed inside a tube possessing optical windows at both ends. A carbon arc light source, containing radiations of all wavelengths, is optically focused to pass through the gas-containing tube and impinge on a slit, behind which is placed a photographic plate. The elements present in the gas sample absorb some of the light source, and this produces dark lines on the photographic plate when it is developed. These lines are characteristic of the elements present, and the density of the lines proportional to the amount of the element present in the gas sample.

## 3. Chemical detector tube methods

Many industrial companies have developed and marketed a system by which gases, usually one gas or vapour, can be estimated, with about 10% accuracy, by a chemical reaction that produces a specific colour or stain in a glass reaction tube. The Draeger Company (Drägerwerk AG, Lubeck) pioneered this method and market a comprehensive range of detector tubes for 380 different gases and vapours. Other British and Japanese companies have modified the Draeger system, but in essence the same chemical reactions are used. The main advantages of this method are that it requires a simple fixed volume air pump to pass a known volume of air through the chemical reaction tube, and the cost of each analysis is about £1.50 and can be carried out by semi-skilled staff. The disadvantages are that a separate tube is required for each measurement made (in some cases a range of tubes is required to cover high and low concentrations of a particular gas), the analysis only relates to a single 'snap sample' of air taken at the time and the accuracy of the method is not high (10–15%). However, many workers find that this system is very useful particularly for on-site safety staff in assessing whether the occupational exposure limit for a particular gas has or has not been exceeded.

## 4. Chemisorption methods

It has been known for a long time that gases and vapours adsorb onto granules, powders and even clothing. Oxygen gas adsorption has caused serious burns and even death to gas welding workers. Adsorption of gases and vapours can be either physical adsorption or chemisorption, or a combination of both. With chemisorption, the gas combines chemically with the absorption matrix and the binding can be quite strong, effectively irreversible at room temperature. Desorption of the gas requires the application of heat or chemicals to the absorption matrix. Physical adsorption of gases is less tenacious. These principles are now used in the commercially available gas diffusion samplers, such as the Orsa system of Draeger Aktiengesellschaft of Lubeck. The method depends on the gas or vapour diffusion into a tube sampler containing activated charcoal. The gas readily absorbs onto the charcoal and is firmly held in the matrix. More than one gas or vapour can be adsorbed so that mixed atmospheres can be monitored. The amount of gas or vapour contaminants is a function of the concentration present, the diffusion coefficient of the individual vapour, as well as the exposure time (usually 8 hours), and the device constant. The tube sampler having been exposed can be sealed and sent to an analytical laboratory where the gas or vapour from the tube can be desorbed and analysed by gas–liquid chromatography. A number of toxic gases and vapours can be accurately estimated by this method. The accuracy of the measurements including the desorption from the adsorbent filling is about 95%. The method offers additional advantages to other methods in that a number of different vapour components can be sampled together, and each of these components can be identified and measured by gas–liquid chromatography. The tube sampler lends itself to use as a personal breathing zone monitor.

## 5. Gas–liquid chromatography

Mixtures of gases or volatile liquids or vapours can be separated and estimated using gas–liquid chromatographic methods. The technique involves placing a gas or liquid sample into a stream of inert gas, e.g. argon or nitrogen, which is passing through a metal or glass column packed with a heated inert solid support coated with a high temperature resisting oil or polymer. The mobile gas phase passes through the column and its components separate out depending on their partition coefficients between the gas and the stationary phase on the support. The separated gases can be identified and measured by passing them through a detector, e.g. flame ionization type, or a mass spectrometer.

## 6. Infra-red and ultraviolet absorption methods

In the last ten years, instruments have been produced that measure the concentration of gases present in the air or vapour mixtures. The methods are based on the amount of ultraviolet or infra-red absorption occurring when a vapour mixture containing the gas to be measured is passed through and absorption tube of known length. Many of the 400 gases and vapours listed by the US Occupational Safety and Health Administration (OSHA) can be determined with considerable accuracy at the ppm level by portable infra-red detectors, such as the Miran series of instruments (e.g. Foxboro Analytical Ltd.); instruments of this type have monochromators which enable the operator to change the wavelength of the instrument, usually from 2 to 20 µm and thus to change analysis from one gas or vapour to another. These instruments, although expensive, offer an accurate and versatile method of determining gas concentrations, particularly at low levels in a gas mixture. The only drawback to these methods is the resolution of gases with similar absorption wavelengths. Atomic absorption spectrophotometers are used for metal vapours such as mercury whose atomic vapour absorbs strongly at 253.6 nm and this feature enables mercury concentrations in air to be measured well below the OES of $25 \mu g/m^3$ in air. These analytical methods can be operated as continuous flow through cell systems, enabling continuous recording of the instruments, output to be made. Modern process control and safety systems make use of these detection methods.

## 7. Paramagnetic methods

Gases, such as oxygen, can be determined quickly and accurately by making use of their paramagnetic properties, produced by the presence of unpaired electrons in the atom's outer electronic shell. Air or gas containing oxygen can be passed through a cell which forms part of a sensitive magnetic balance. The changes in the field produced by the paramagnetic gas is proportional to the amount of gas in the cell. This method of detection is extremely sensitive for oxygen since 1 mole of oxygen contains $10^{23}$ molecules of which approximately $10^{15}$ of them have unpaired electrons. The theoretical sensitivity of the method should be 1 part of oxygen in $10^8$ parts of other gases or a detection limit of $10^{-8}$ grams of oxygen. Methods involving the flow of gas through a detection cell offers the opportunity of continuous monitoring. This technique would be most suitable for measuring oxygen concentrations present in gas mixtures used in industrial processes.

## 8. Polarographic methods

Gases can pass through thin semi-permeable membranes which retain liquids. In the polarographic method, a thin liquid film of electrolyte is maintained between two electrodes by means of a gas permeable membrane which together form an electrolysis cell. The negative electrode is usually a platinum wire, the positive electrode consisting of a silver wire or strip coated with a fine deposit of silver chloride. Other metals and metal ion combinations have been used. The potential difference across the cell is maintained at the decomposition potential of the gaseous ion being considered. Under these conditions, the current produced when the charged ion is attracted to the oppositely charged electrode and gives up its charge, is proportional to the concentration of the gas present. In the case of oxygen, the potential difference across the cell to equal the decomposition potential is 0.61 volts with respect to hydrogen. This type of polarographic cell can be placed in a gas-filled enclosure, in a stream of a gas mixture, or in aqueous solution in which the gas has dissolved. The method has applications for process control in industry, in environmental monitoring of rivers and water supplies, and with suitable modifications, e.g. as needle-shaped micro cells capable of measuring the oxygen concentration in the bloodstream of living subjects.

# PART 2
## Safety precautions and planning

All the gases presented in this book possess some degree of risk and hazard. The purpose of this part is to list the safety precautions and planning that are necessary before any work is started.

### Risk assessment

**Before any work is undertaken, a full risk assessment of the proposed procedures must be made**. This means investigating all the risks including physical, chemical (covered largely by a COSHH assessment), biological, and toxicological including occupational health risks to the staff. In addition, special attention must be given to the effects of an accident or explosion in the process causing a spillage of chemicals or rupture of gas lines or storage facilities possibly resulting in fire at the workplace. This preplanning of work is essential if the process to be carried out is on an industrial scale using tonnes of material. In teaching and research laboratories, where the quantities used will be very much smaller, the risk assessment should form part of the general assessment made of laboratory hazards.

**The risk assessment** must identify the hazards to health and safety to which employees are exposed while at work or any other persons who may be exposed as a result of the work, and specify the measures that will be taken to minimize the risks to a level acceptable by current legislation or good working practice. For example, if the process being undertaken involves the use of a toxic gas, the employer is required to consider the nature of the hazards involved in using the gas, its storage arrangements in relation to the workplace, other buildings and processes, the movement of gases either by cylinders or pressurized pipelines from a liquefied gas reservoir, and safety procedures in the event of an accidental release of the toxic material.

Where a flammable gas is used in large quantities, e.g. oxygen in a hospital or a large steel fabricating works, safety procedures must be planned to deal with leakage from the pipe gas system, and the isolation of the pipe system in case of fire or line rupture. For example, many large hospitals supply medical grade oxygen to wards and operating theatres using a pressurized pipeline system supplied from a large liquid oxygen storage tank. Oxygen is drawn from the storage tank through the pipework to the wards via a series of non-return isolation valves which can shut off the oxygen supply to a particular ward in the event of fire, while maintaining the supply of oxygen to the rest of the hospital.

Risk assessment must also consider the relative toxicity of the gas used and any products that may be formed from it. Phosgene, a highly poisonous gas, can be easily formed from processes where chloroform or similar organohalogens are heated. If toxic gases are or may be released from chemical processes, the workplace will require adequate mechanical ventilation that removes these gases from the workplace and either passes them through an appropriate filter system that removes the gas, or if the gases are of low toxicity, the extract system dilutes the emerging exhaust so that the concentration of these contaminants is less than the occupational exposure standard for the contaminants. This is particularly important if the intended chemical process is to be carried out near to a residential area.

**A complete risk assessment consists of a written statement explaining the risks and the procedures by which these risks can be eliminated or reduced to acceptable limits. This document must form part of the employer's safety policy and local rules governing the work, and should be available to all staff who have a need to know its contents.**

## Fire safety
**Fire safety arrangements are usually based on the two tier structure.**

**At the first level**, the workplace will have a fire prevention policy in which the fire risks are assessed, with the advice of a trained Fire Prevention Officer from the local fire brigade, and then arrangements are made to ensure the early detection and reporting of fire and its isolation. Management will be expected to provide basic fire safety training for all staff, arrange fire drills and provide special training for staff who will act as fire wardens.

**At the second level** is the professional fire brigade firefighters who are trained to fight, contain and extinguish the fire, rescue any trapped persons, and ensure the premises are safe. They are also able to deal with toxic chemical releases, but in some situations, particularly in the chemical industry, would require information from local staff as to the nature and location of stored chemicals or gas cylinders.

**The extent of the fire safety arrangements** in a workshop or laboratory will depend upon the nature and quantity of the gases or substances used, and the location of the workplace. In urban areas, the local fire service will have estimated the potential risk, and entered these in a central risk register usually from information obtained at site visits by the Fire Prevention Officer and/or firefighters from the local station who would respond to any fire calls from the premises. On receiving a fire call from a particular workplace, the brigade will know the nature and extent of the risk usually written on a computer printout before they leave the fire station.

### Works fire teams
Where the works or laboratory is situated in a rural area, the attendance of the nearest fire appliances may take some time. Under these circumstances, consideration should be made to providing a local trained works fire team which can tackle the fire and its toxic effects prior to the arrival of the brigade.

Local fire teams are also very desirable in factories where gases or vapours used in processes are particularly toxic, e.g. in the production of pure nickel using nickel carbonyl. Fire teams must be properly and regularly trained. They require modern British Standard approved equipment including protective clothing and breathing apparatus. Their principal duty is to save life and contain the fire, if possible, with no danger to themselves, using available fire fighting equipment until professional firefighters arrive to take charge. The standard of training for these factory teams should be not less than that received by the part-time retained brigade firefighters.

## Portable fire equipment
This equipment is designed to be used by the staff who discover a fire. The choice of the correct type of extinguisher must take into account the type of flammable material in the area.

1. **Water (gas cartridge type) extinguishers, coloured red**, are suitable for quenching a wood or paper material fire. The extinguisher consists of a metal cannister filled with 10 litres of water which can be directed onto the base of the fire via a short rubber hose. The water is expelled from the extinguisher by operating a handle which releases carbon dioxide gas from a cartridge held within the extinguisher.

Where there are no contraindications, such as electrical apparatus or non-miscible organic liquids on fire, water provides a good extinguishing agent. More recently, water extinguishers have been produced which contain a light foaming agent known as AFFF which enables them to be used on flammable liquid and electrical fires. These fire extinguishers are extremely effective.

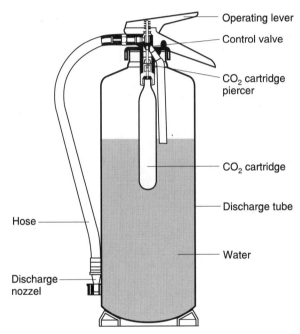

**Figure 1**   Water extinguisher (gas cartridge type)

2. **Carbon dioxide extinguishers, coloured black,** are used where there is a need to protect valuable electrical equipment, or there are chemicals present in the area which exclude the use of water. These extinguishers produce a cloud of carbon dioxide gas and solid $CO_2$ particles when operated, which cool the fire and blanket it with the heavy carbon dioxide, starving the fire of oxygen.

Carbon dioxide should not be used as a fire fighting agent in basement or confined spaces. The build-up of high concentrations of carbon dioxide can be lethal to personnel fighting or those escaping from fire. Carbon dioxide must not be used on burning metal fires.

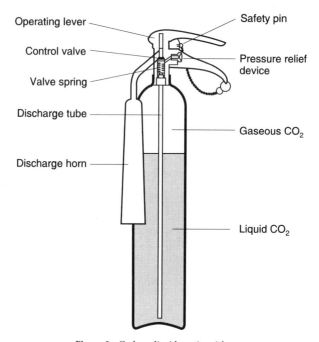

**Figure 2**  Carbon dioxide extinguisher

3. **Halon (bromochlorofluoromethane BCF type) extinguishers, coloured green,** are used in factories, laboratories and workshops where there is a possibility of oil and flammable spirit fires. The BCF agent was originally developed to quickly extinguish fires in aircraft. Operating the extinguisher releases a stream of vaporizing liquid and gas over the fire, which effectively covers the fire and excludes oxygen from it in a similar way to carbon dioxide. However, the presence of the bromine atom in the BCF molecule acts as a chain terminator in the oxidation process of the fire. One disadvantage in using this agent is its unsuitability for dealing with fires involving hot metal surfaces. The BCF molecule is thermally degraded to produce hydrogen halide gases which, of course, are toxic and corrosive. Because of this, the use in confined spaces requires adequate ventilation of the area before reoccupation by staff. Recently, this extinguisher has been the subject of adverse publicity from the environmental lobby due to their view that the gas might affect the ozone layer and contribute to global warming. These extinguishers are not suitable for laboratories or workshops containing valuable electronic apparatus.

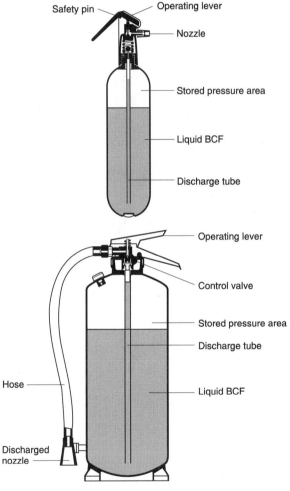

**Figure 3** Halon extinguisher (BCF type). Upper drawing shows the small 0.7 kg size. Lower diagram shows the larger 7 kg size

4. **Powder extinguishers (gas cartridge type), coloured blue**, contain of a mixture of dry inert powder such as fine silica mixed with solid sodium bicarbonate powder. The powder is propelled from the extinguisher with the aid of carbon dioxide gas from a gas cartridge. These extinguishers are very effective in covering the fire with a powder blanket which releases carbon dioxide and excluding oxygen from the fire. In factories where valuable machinery is sited, the use of this type of extinguisher might not be desirable, as considerable damage will be done to machine bearings or delicate instruments if they are covered with fine abrasive powder.

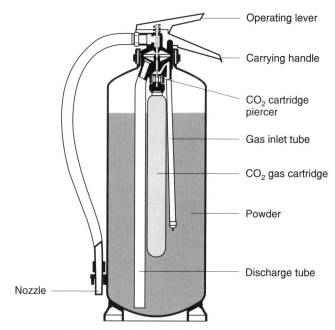

Operating lever

Carrying handle

$CO_2$ cartridge piercer

Gas inlet tube

$CO_2$ gas cartridge

Powder

Discharge tube

Nozzle

**Figure 4**   Powder extinguisher (gas cartridge type)

5. **Foam extinguishers (gas cartridge type), coloured cream**, consist of a cylinder containing chemical reagents that when mixed together produce a stable foam which is discharged through a cone nozzle propelled by means of carbon dioxide from a gas charge. The foam forms a blanket over the fire, smothering the flames and excluding oxygen from it. Foam fire extinguishers are used to deal with oil or grease fires where the use of water is contraindicated. In many locations, the foam extinguisher has been superseded by the water/AFFF type.

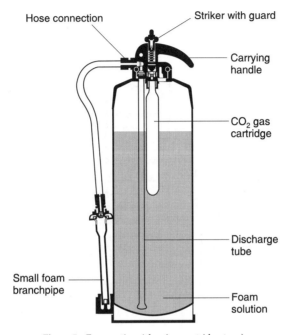

**Figure 5**   Foam extinguisher (gas cartridge type)

6. **Glass fibre fire blankets** are very useful in the event of fires involving small open vessels or small gas cylinders where the fire is in the cylinder reducing-valve. If the blanket can be placed over the vessel or cylinder in the early stages of the fire, portable extinguishers can be directed onto the fire blanket reducing the severity of the fire and allowing staff to evacuate safely.

## Fixed fire equipment

1. **First aid hose reels, painted red.** In many modern buildings, 1 inch first aid hose reels are provided in corridors or near external storage areas, which are connected to the water main and when operated deliver a jet or fine water spray. These hose reels can be used by trained staff to provide a more substantial back-up to the water/gas expelled fire extinguisher prior to the arrival of the fire brigade. Blow-back fires in acetylene cylinders can be controlled by directing a water spray over the cylinder, from a safe distance, using a first aid hose reel, in order to cool the cylinder and prevent an explosion.

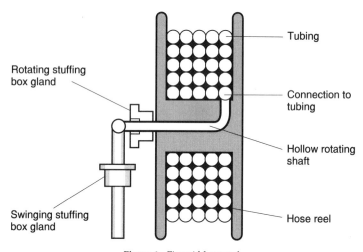

**Figure 6**   First aid hose reel

2. **Fixed sprinklers and gas drenches.** In factory installations where there is a possibility of a gas leak or fire, e.g. in nitric acid production, the plant is situated in a open area contained in a concrete bund and fitted with a fixed water sprinkler system. If a failure occurs, the whole plant is enveloped in a fine water spray which removes all the water soluble toxic gases. Other processes using non-water miscible organic gases or liquids may require the provision of a carbon dioxide or halon gas drench which is activated by the presence of fire or smoke.

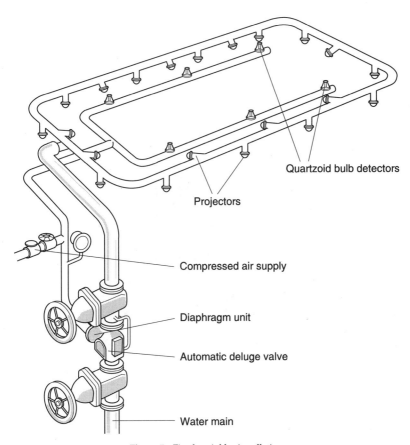

Quartzoid bulb detectors

Projectors

Compressed air supply

Diaphragm unit

Automatic deluge valve

Water main

**Figure 7**  Fixed sprinkler installation

3. **Dry riser installations.** In many modern buildings and factories, dry risers are often installed during the construction of the building. The dry riser consists of a network of metal pipes laid throughout the building so that the standard fire brigade branch hose can be attached at any dry riser couplings fitted at convenient positions throughout the building. If fire breaks out in the building, the attending fire appliance can connect its fire pump outlet to the dry riser inlet, usually situated at ground floor level. This enables the whole riser system to become charged with water under pressure so that the firefighters can operate at any point in the building without the need for laying extensive hose lines.

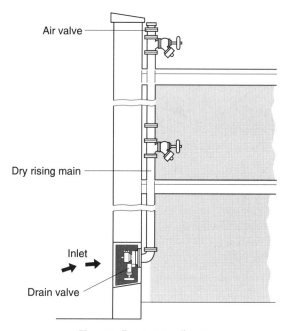

**Figure 8** Dry riser installation

# PART 3

# First aid and medical treatment of injured or exposed persons

Throughout this book, advice is given for first aid treatment of casualties involved in incidence with toxic material and gases. The advice has been restricted to basic first aid, to ensure that the victim is removed as safely and quickly as possible from the contaminated area, and is then treated to ensure breathing and circulation are supported until professional medical or paramedical help arrives.

### First aid training policy

Current first aid training policy specifies the number of trained first aid staff required for any factory, workshop or laboratory. The number of trained and available first aid staff, and the facilities that may be needed in any work situation where toxic materials are handled should be discussed with safety experts and the employees. The standard of training must be not less than that approved by the HSE. Special training will be required for first aid persons who are intending to treat staff exposed to special known hazards. The use of antidotes or toxic inhibitors must be included in specialized training programmes. Enhanced facilities will be required in workplaces that are situated in remote areas, or where the attendance of paramedical or medically qualified staff could be delayed. In many workshops a minimum of one trained person per 50 employees may be insufficient to provide good first aid cover at work. The numbers trained should reflect the scale and nature of the work undertaken.

### First aid training

Training for the first aider must include knowing when and how to use breathing apparatus, and what type of equipment is most suitable. Laboratories where toxic gases are used must have appropriate breathing equipment available near to the workplace to enable victims to be removed from the contaminated area without serious risk to rescuers. Removal of unconscious victims should only be attempted by staff working in pairs. The availability of an oxygen-enriched resuscitator or an oxygen supply mask is essential.

### Provision of first aid equipment

Portable first aid kits are available from suppliers, which contain the basic equipment required to render first aid to injured staff. It is important to ensure that any first aid kit contains dressings that can cope with accidents involving large wounds, e.g. those similar to army field dressings, pressure bandages, antiseptic wound cleaning wipes, and an airway which can be used to assist mouth-to-mouth resuscitation fitted with a one-way air valve to prevent expired air from the patient entering the first aider's lungs. The small office type first aid kit is not suitable for workshop or laboratory accidents.

### First aid treatment
In first aid treatment, the procedures should be carried out simply, efficiently and without delay. With the data provided for each gas in this book, the first aid treatment has been divided into four parts relating to: (a) inhalation; (b) ingestion; (c) skin contamination; and (d) eye contact with the gas. In each case, the procedures are designed to remove or dilute the damaging effects of the chemical by using simple ready to hand facilities. Often the effects of inhalation of a gas can be reduced by removing the victim from the contaminated area, and allowing the person to breathe fresh air.

### First aid action following rescue
Having removed the victim from the accident scene, the first aider must make a quick assessment of the accident priorities. Restoration of normal heart action and breathing must take precedence over minor bleeding or wounds. If the victim is unconscious and has severe burns, skin or eye contamination with the toxic material, it will be necessary to organize a small team of staff who will deal with these problems without delay. It is essential in the safety preplanning that such arrangements are made in advance so that the casualty can be treated in the most speedy and effective way.

### Resuscitation and the restoration of normal breathing and circulation
The first essential following the rescue of a person(s) from a contaminated area is to establish normal breathing and blood circulation. This is known as the **ABC** of resuscitation, i.e. Airway, Breathing, and Circulation.

If more than one victim is involved, resuscitation priorities must be established.

1st Priority  –  Casualties who are unconscious with no pulse and/or are not breathing.
2nd Priority  –  Victims who are unconscious but breathing.
3rd Priority  –  Casualties who are conscious and breathing but suffering from gas inhalation.

## Airway

The opening of the airway is the first priority for an unconscious person.

1. If the mouth has vomit, or any material obstructing the airway such as false teeth, this should be removed using a handkerchief or cloth.

2. Place the casualty on his/her back and flat on the ground, then move the lower jaw forward tilting the head backwards so that the position of the chin is vertical. This will lift the tongue away from the back of the throat and open the airway. **You can place a small soft object under the neck to help maintain the head position but not under the head**.

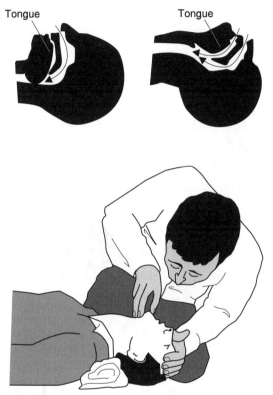

**Figure 9** Upper diagram shows the head tilt-lift method to open an airway and prevent obstruction by the tongue. Lower diagram illustrates the position of the hands of the operator for the head tilt-chin lift method

3. When the airway is clear, the casualty may start breathing normally. If this occurs, place him/her in the recovery position (Figure 13). If not, start to perform mouth-to-mouth resuscitation preferably using a non-return plastic airway or resuscitation bag and mask (Figure 10).

## Mouth-to-mouth resuscitation

1. With the airway opened as described, pinch the victim's nose, then while holding the victim's mouth wide open, the rescuer should take a deep breath and, opening his/her mouth wide, make a seal around the patient's mouth and blow in air. The patient's chest should rise.

2. Repeat with three further quick breaths. The rescuer should remove his/her mouth from the victim and allow air to escape from the lungs. This procedure should be repeated 12 times per minute.

3. If the victim is not breathing but has a pulse, continue this treatment until paramedical aid arrives.

4. If the patient is not breathing and has no perceptible pulse, external chest compression must be carried out together with mouth-to-mouth resuscitation.

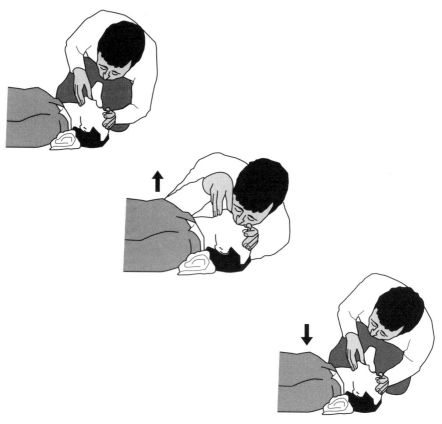

**Figure 10**   Mouth-to-mouth resuscitation. Upper diagram shows positions of hands. Middle diagram shows inflation of the lungs. Lower diagram shows rescuer observing lungs deflating

## External chest compression
**This should only be attempted if there is no heart beat or carotid pulse.**

1. The casualty must be placed on his/her back face upwards on the floor or suitable hard surface.

2. The rescuer should position him/herself by the victim's chest, in line with the heart and place the heel of one hand on the lower half of the breast bone.

3. The other hand of the rescuer should be placed on top, keeping the fingers and palms clear of the ribs.

4. The elbows are then to be straightened, and the shoulders positioned directly over the hands.

5. The rescuer then rocks forward, keeping the arms straight and pressing vertically down on the victim's chest, lifting his arms to allow the victim's breast bone to rise again.

6. This should be repeated every second until the carotid pulse returns to the victim.

If the rescuer has to perform both mouth-to-mouth resuscitation and chest compression, he should give a cycle of 15 chest compressions to 2 mouth-to-mouth resuscitations.

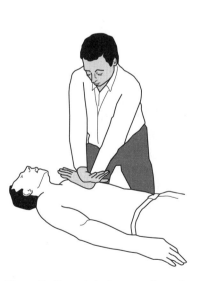

**Figure 11** External chest compression: note the position of the rescuer's arms and body performing the procedure

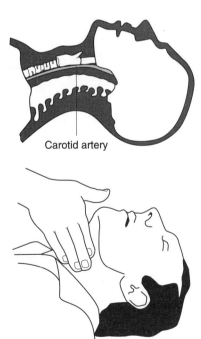

Carotid artery

**Figure 12** Position of the fingers when feeling for the carotid artery pulse in the neck

## The recovery position

Patients who are breathing and have a pulse, but who are still unconscious, must be placed in the recovery position. The position is designed to prevent the patient's tongue from obstructing the airway and preventing the inhalation of any vomit or blood.

1. The victim is placed on the floor or stretcher so that he/she is lying on the side with the face turned towards the uppermost shoulder.

2. Ensure that the mouth is free from dentures or any objects that might obstruct breathing.

3. Loosen clothing, collars and belts, and remove spectacles if worn.

4. Cover the casualty with a light blanket or covering unless he/she has a high body temperature.

5. Give oxygen, or oxygen-enriched air, using a face mask, if available, to aid recovery.

6. **Do not leave the casualty, check frequently that breathing and pulse are normal. If breathing or pulse fails, carry out artificial respiration or chest compression as necessary.**

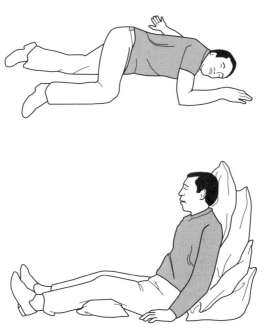

**Figure 13**   Upper diagram shows **the recovery position** used if the casualty is breathing, has a pulse, but is not fully conscious. Lower diagram shows **the sitting-up position** used when lung congestion is suspected

## Problems with mouth-to-mouth resuscitation of casualty overcome by a toxic gas

For persons unconscious as a result of inhalation of a toxic gas the use of mouth-to-mouth resuscitation methods, advocated by most first aid courses, must be viewed with caution. **There is a possibility that the rescuer can become affected by the presence of the toxic gas in the exhaled air from the patient. This is particularly important if the casualty has been affected by cyanide or hydrogen sulphide gas.** The best and safest method of resuscitation in these circumstances is the use of an Ambubag resuscitator preferably fitted with the means to give supplementary oxygen. They are simple to operate, and can be used on the victim during the rescue, if evacuation by stretcher from a contaminated area is protracted. **Under no circumstances should a rescuer remove his/her breathing set mask to give fresh air to the patient during a rescue. Both lives can be lost by this action.**

## Danger from adsorbed gases on clothing

It is important to remember that toxic gas victims may have considerable amounts of the gas adsorbed onto clothing particularly if the person is wearing a thick woollen jacket. This gas can become desorbed from clothing when the casualty is removed into a warm first aid room or ambulance. Where possible the victim's heavy jacket should be removed, and the casualty covered with a warm blanket. If the jacket is required for further examination at the hospital, it should be placed and sealed inside a thick plastic bag.

## Chemical contact with the eyes or skin

With any chemical contact to the skin or eyes, the first golden rule is to wash the affected area with a steady stream of cold tap water, preferably from a hosepipe, for at least 15 minutes. The easiest way to irrigate the eye is to lay the patient on the ground or over a sink and direct a slow stream of cold tap water from a hose so that it fills the eye socket thus bathing the whole eye ball and surrounding tissues with water. Under these circumstances, the patient can open and shut the eye lids to ensure that no toxic chemical is trapped in the recesses of the eye.

Cold water is preferred because it minimizes the absorption of the chemical through the tissues into the body; the cooling effect of the water reduces the dilatation of blood vessels in the surrounding tissues and also reduces the pain caused by the exposure.

Apart from specific antidotes mentioned in the text, first aiders should not attempt to apply medications to wounds, etc. especially in the case of chemical burns. Wounds should be covered with either a dry dressing or a sheet of clingfilm to prevent infection and any follow-up treatment should be given by paramedical or medically qualified staff.

## Inhalation of volatile chemicals and gases

Persons who have been affected by chemical fumes or gases must be removed as soon as possible to a safe uncontaminated area for first aid treatment. If the victim is conscious, breathing and with good circulation but affected by gas inhalation, they should be kept calm and at rest, preferably lying down with the upper part of the body raised to aid breathing. Cover with a light blanket to keep the patient warm but not hot. Oxygen or oxygen-enriched air from a face mask, if available, should be given at intervals as required. The greatest danger for gas victims is the development of fluid filling the lungs (pulmonary oedema). Any victim of gassing must be seen by a paramedic or doctor and transferred to hospital as soon as possible.

# PART 4
## Personal safety equipment

Work with toxic chemicals requires the wearing of appropriate personal protective clothing.

### Safety clothing
In industry, workers must be provided with chemical resistant overalls or coats with leggings, and safety boots or shoes. The design of the clothing should offer full protection to the whole or part of the body if liquid chemicals are spilt. In laboratories, laboratory type protective coats should be worn. Tests show that polyester/cotton material gives superior protection to the body from chemical contamination than the standard cotton drill cloth. The latter material absorbs large amounts of liquid when spilt onto its surface. The wrap over front fastening reduces the risk of contamination of personal clothing worn underneath.

In hazardous locations, where there is a risk of fire and explosion, staff should be discouraged from wearing nylon clothing or underwear. In dry conditions, this material can easily generate static electricity leading to spark discharges. If the external clothing catches fire, the nylon undergarments can melt and glue onto the skin making removal extremely difficult and painful.

### Hand protection
For handling corrosive gases and liquids, thick nitrile gloves or gauntlets should be worn. Where liquefied gases are handled or poured from containers, thick leather loose fitting gloves or gauntlets are needed to protect the hands. They must be loose fitting to enable easy removal in the event of cryogenic liquid entering the inside of the gauntlet. The choice of gloves, etc. must be made in relation to the chemical reactivity of the glove material to the gas or vapour concerned, e.g., rubber gloves are rapidly destroyed by contact with liquid bromine or its vapour, so these gloves would offer no protection to the hands in that situation.

### Eye protection
**Eye protection should be everyone's obsession.** The maxim that 'skin regenerates but eyes don't' should be borne in mind by everyone at work. **You only require one error without eye protection to lose your sight.**

Manufacturers produce a whole range of British Standard approved eye protection ranging from simple eye shields, safety glasses, face shields designed to protect against the impact of flying objects, and, in some cases such as welding shields and goggles, also to give infra-red and ultraviolet light protection.

23

In factories, workshops and laboratories, eye protection should be part of standard safety practice for all staff throughout the workplace. Notices to that effect should be displayed. The eye protection policy must apply to all staff, even visiting managers and VIPs. Consultants in hospitals thought, at one time, that they had divine protection from accidents and pathogens, but, to their credit, have discovered that they can be injured or affected like other mere mortals.

Prescription safety glasses should be worn in preference to contact lenses in all situations where chemicals could accidentally enter the eye. Water irrigation of the eye during first aid treatment would not remove any toxic chemicals trapped between the cornea and the contact lens, and permanent damage to the eye could result.

## Respiratory protection

Respiratory protection devices exist in a number of forms:

(a) **Simple paper/cloth filter masks** which are used to remove low toxicity nuisance dusts or particles from entering the lungs. They are meant to be used once for short duration work periods.

(b) **Half and full faced filter masks** usually fitted with a pre-filter and a filter cartridge containing absorbent material which will adsorb or trap a specified dust, fume or gas. These masks are designed to be used in atmospheres where the maximum concentration of the contaminant is known and where there is no deficiency of oxygen present in the air. The filters fitted to these masks are colour coded for the type of gas or fumes they trap and they have a specific working life measured in hours. The disadvantage of these filter gas masks is ensuring the filter is in good order and not heavily loaded with adsorbed gas or fumes. This could cause breakthrough of the toxic gas into the mask. The best practice is to replace the filter after each work period especially if the work requires protection from highly toxic volatile chemicals. Many manufactures now make gas masks in which the gas adsorption filter is in the form of a spin-on threaded canister. This enables a range of canisters, protecting from different gases, to be used on a standard gas mask.

(c) **Compressed air breathing apparatus (CABA).** This type of respiratory protection is needed when highly toxic gases or vapours are involved and/or when the oxygen concentration in the air is unknown or thought to be deficient for normal breathing. **CABA cannot be used by untrained persons, nor should it be used in a rescue attempt by a single individual. Entering a toxic gas environment is a dangerous operation requiring a trained team of rescuers.** The equipment also requires regular expert maintenance. Advice should be sought from the local Fire Authority on the desirability of setting up such a CABA rescue team.

(d) **Personal CABA sets with a 10 minute air cylinder** are sometimes sited in hazardous areas to assist the escape of staff following an accident or fire. **These small sets are for escape purposes only and are not to be used to rescue other persons.**

(e) **Breathing air lines with pressurized suits** are normally used in industrial plants where staff work for long periods in contaminated areas. The pressurized suits are made of chemical resistant material often in the form of a fully covered overall with built-in visor and having gas tight seals. Breathing air is supplied to the suit by means of an armoured air line to ensure positive air pressure inside the suit and prevent any external gases entering it. Some suits are fitted with compressed air cylinders, in addition to the air line, as an extra precaution, or for short duration work away from the air-line supply. Suits of this type would be required for work on toxic gas lines in a factory producing nickel carbonyl.

# PART 5
## Occupational health

Occupational health is an essential part of the safety structure of any workplace. Legally it is part of Health and Safety Legislation which embraces the Common Law Duty of Care and the more specific Statutory Instruments and regulations relating to specific dangerous materials, e.g. lead. The purpose of occupational health in the workplace is to maintain and promote the physical and mental well-being of employees. Good occupational health facilities would aim to provide workers with a comprehensive range of services including:

1. **Pre-employment medical examinations for all staff** graded in extent by the nature and potential health risks of the work to be undertaken. This is particularly important for new staff who may have non-occupational diseases or handicaps such as diabetes or heart disease.

2. **Operation and control of occupational hygiene.** The hygiene provisions should include the general cleanness and housekeeping of the workplace, its supervision by trained staff, the prevention of occupational disease by monitoring of, and if necessary arranging for, the upgrading of the working environment, and the supervision of personal protective equipment issued to exposed workers.

3. **Advice to management and to worker's representatives** on all aspects of the working environment, including prevention of accidents and occupational diseases.

4. **Health education and training,** including general and specialized first aid related to the particular risks associated with the workplace.

5. **Record keeping,** including the analysis and recording of absentees due to accidents, sickness and occupational diseases. The preparation of reports concerning the health status of the staff and occupational hygiene of the workplace.

6. **Establishing a good working relationship** between the occupational health services and the other branches of the organization.

The size and extent of the occupational health services in any workplace will depend upon the number of staff employed, and the nature of the work being carried out.

To assist in the planning of occupational health, a section has been allocated for each gas or fume, and covers the occupational hazards of chronic and acute exposure (EOHS, pp. 1504–1528 gives detailed information on the organization, set-up and training of occupational health and hygiene facilities indicated here).

# PART 6

Health and safety data on gases and fumes

# Acetylene  $C_2H_2$

## Exposure limits

No OES or STEL quoted because acetylene is relatively non-toxic but can act as an asphyxiant. EOHS suggest that the maximum permissible working limit for acetylene should be less than 5000 ppm (0.5%) in carbide generated acetylene.

## Description

A colourless gas with a garlic like odour when supplied as a commercial grade gas. **Highly flammable and explosive when compressed.**

## Properties

The pure acetylene has a faint odour of ether but the commercial grade gas supplied in cylinders has phosphine, hydrogen sulphide and ammonia as impurities giving the characteristic garlic smell. Acetylene is soluble in water and organic solvents.

MW = 26,   SG = 0.62 (–82°C),   MP = –81.8°C,   BP = –75°C,
VD = 0.9,   VP = 40 atmos. ($4052 \times 10^3$ Pa) at 16.8°C,   FP = –32°C,
EL = 2.5–81%,   IT = 305°C.

## Detection methods

Flammable gas monitor.
Chemical reaction tubes.
Small leaks can be detected by its characteristic smell.

## Precautions when using the gas

Because the gas forms an explosive mixture with air at concentrations of gas from 2.5 to 81%, **fire and explosion represents the greatest risk when using this gas.** Fortunately, the characteristic garlic smell of acetylene allows small leakages to be detected and located by means of an application of a soap solution onto the pipe or joint. **Never use a match or flame for this purpose.**

Acetylene is supplied in cylinders holding 4–8 m$^3$ of gas dissolved in acetone impregnated material inside the cylinder. **Because acetylene becomes very explosive when heated or compressed, cylinders containing acetylene must be stored upright and in cool, well-ventilated areas, protected from direct sunlight.** The cylinders should be handled with care, and not dropped when being transported. It is important to fit anti-flashback devices to acetylene cylinders to prevent internal fires in cylinders. These can occur when acetylene is used in connection with oxy-acetylene cutting equipment. **Never allow acetylene to come into contact with copper pipework as explosive copper acetylide is readily formed.**

31

## Occupational health

Normal working with acetylene does not constitute a health hazard as the gas is relatively non-toxic. Accidents at the workplace involving rupture of pipework releasing large amounts of acetylene could produce intoxicating and asphyxiating atmospheres. Concentrations of gas of less than 10% produce slight intoxication, but if the level rises to 35% the victim could be rendered unconscious in 5 minutes.

---

**First aid**

Inhalation

Persons who have inhaled acetylene gas should be removed from the contaminated area into fresh air as soon as possible, and allowed to recover. **If the acetylene atmosphere is contaminated with phosphine from a calcium carbide generator, or the patient is unconscious, oxygen-supplemented artificial respiration should be given, the subject kept warm and at rest, and treatment for phosphine poisoning considered. Medical opinion should be sought without delay.**

Skin and eye contact

Skin or eyes affected with acetylene gas or liquid should be irrigated with a stream of cold tap water for at least 10 minutes to remove all traces of acetylene. **In the case of eye contact or serious skin contamination, the patient should be referred to hospital without delay for specialist medical treatment.**

---

# Alkylamines

| | |
|---|---|
| Methylamine | $CH_3NH_2$ |
| Dimethylamine | $(CH_3)_2NH$ |
| Ethylamine | $(CH_3CH_2)NH_2$ |

**Exposure limits**

| | | |
|---|---|---|
| Methylamine | OES = | 10 ppm (12 mg/m³) |
| Dimethylamine | OES = | 10 ppm (18 mg/m³) |
| Ethylamine | OES = | 10 ppm (18 mg/m³) |

## Description

The most common alkylamines are the mono, di, and trimethyl derivatives of ammonia and ethylamine. **They are all gases at room temperature, heavier than air and possessing a very alkaline reaction. Alkylamines are flammably explosive when mixed with air.** These gases have a fishy ammoniacal smell.

## Properties

**Methylamine – a colourless gas, very soluble in water**

MW = 31.1, SG = 0.70, MP = –93.5°C, BP = –6.3°C, VD = 1.1,
VP = 1520 mmHg (202.2 × 10³ Pa) at 25°C, FP = 0°C,
EL = 4.9–20.7%, IT = 430°C.

**Dimethylamine – a colourless gas, very soluble in water**

MW = 45, SG = 0.68 (liquid), MP = –96°C, BP = 7.4°C, VD = 1.6,
VP = 1520 mmHg (202.2 × 10³ Pa), FP = 12.2°C, EL = 2.8–14.4%.

**Ethylamine – a colourless volatile liquid with a strong ammoniacal smell**

MW = 45, SG = 0.69, MP = –84°C, BP = 16.6°C, VD = 1.6,
VP = 760 mmHg (101.1 × 10³ Pa) at 16.6°C, FP = 0°C, EL = 3.5–14%.

## Detection methods

Infra-red gas analyser.
Chemical reaction tube.

## Precautions when using the gas

All alkylamines are bases that can form very strong alkaline solutions. It is important that any aqueous liquids formed from the gas should not contaminate the skin or the eyes. **Great care must be taken to prevent spillage of liquid amines onto the body, particularly in the eyes. Protective clothing, full face and hand protection are essential when dealing with these compounds. Cold burns can result from exposure to liquid amines.** The vapour of the alkyamines is easily detected by smell even at low concentrations of the vapour in air. Remember, olfactory fatigue can occur with prolonged exposure to low concentrations.

## Occupational health

These aliphatic amines have not been shown to have any specific toxic properties but occupational health experts are concerned as to the possibilities of ingested aliphatic amines reacting with nitrates or nitrites in the body to form nitroso compounds which are known to be carcinogenic in animals.

Concentrations of amine vapour greater than 100 ppm can lead to coughing due to lung irritation. High concentrations of the vapour may produce severe respiratory distress. Prolonged skin contact with liquid containing the amine can produce skin irritation leading to dermatitis.

---

**First aid**

Inhalation of the gas
Workers overcome by the gas **should be removed to fresh air at once. If breathing is slow or stops, give artificial respiration using an oxygen resuscitator.** The patient should be kept warm and at rest. **Medically qualified assistance is required to determine the seriousness of the patient's condition.**

Ingestion or contact with the nose or throat
Irrigate the nose and throat with cold tap water for at least 10 minutes. **If the patient is able to swallow, encourage them to drink a litre of 2% aqueous citric acid or lemonade to reduce the effect of the alkalinity of the amine.**

Skin contact
**The most serious problem occurs if liquid amine is splashed onto the skin. Remove all affected clothing. Wash area affected with large volumes of cold water, e.g. use a 12 mm flexible hosepipe for 15 minutes.** The affected skin area if unbroken can then be treated with 2% acetic acid or vinegar solution. **Do not cover the skin with a first aid dressing.**

Eye contact
**Efficient first aid is essential to save the patient's sight if liquid amine is splashed into the eye. Immediately irrigate the eye with a slow stream of cold tap water for at least 10 minutes ensuring the water irrigates the whole of the globe of the eye.** (Note: a slow stream of water will be tolerated by the patient and will reduce the possibility of the amine liquid being driven into the deep recesses of the eye's orbit.) **Expert medically qualified opinion is vital and the patient should be removed to hospital without delay.**

---

# Ammonia NH$_3$

**Exposure limits**
OES = 25 ppm (18 mg/m³)
STEL = 35 ppm (27 mg/m³)

## Description
Colourless, easily liquefied, strongly alkaline gas, with a pungent ammoniacal smell, detectable at 20–50 ppm in air. High concentrations of ammonia gas or liquid are very chemically reactive.

## Properties
The gas is soluble in water, ethyl alcohol and organic solvents. High concentrations of ammonia gas or liquid ammonia are very chemically reactive.

MW = 17,  SG = 0.77,
MP = –77.7°C,  BP = –33.3°C.

## Precautions when using the gas
**In laboratories and workshops, the use of 0.88 SG ammonia solution or anhydrous ammonia should be considered hazardous.**

Staff handling ammonia gas cylinders or solutions must wear eye safety shields or goggles. The work should be confined to well-ventilated laboratories in case of spillage, etc. Accidents with ammonia gas cylinders where the valve fails to shut can be dealt with by placing the cylinder into a sink or tank of cold water. In the case of industrial plant using ammonia, the scale of precautions must be considerably greater. All staff must be trained in the handling and use of safety equipment which includes clothing designed to resist an alkaline chemical spillage. Provision must be made for eyewash and shower points, and sufficient self-contained breathing apparatus to rescue any workers affected by ammonia gas or liquid.

## Occupational health
Ammonia is not a systemic poison, but gaseous or liquid ammonia can cause damage to the eyes and the upper respiratory tract, principally by dissolving in the fluid covering the tissues and producing a strong alkaline reaction. Liquid ammonia spilt onto the skin can also cause a cold burn.

### Acute effects
Ammonia gas concentrations of greater than 400 ppm usually produce irritation of the nose and throat and pronounced watering of the eyes. At higher concentrations,

a persistant cough is produced leading to pulmonary oedema and, at very high concentrations of the gas (>5000 ppm), death.

## Chronic effects

Chronic irritation of the eyes, nose and upper respiratory tract may result from exposure to high concentrations of ammonia vapour.

---

**First aid**

If the patient is conscious but coughing, remove from the contaminated area to fresh air and supplement with medical oxygen. If the patient is overcome with ammonia gas, the rescuer must wear breathing apparatus and remove the patient to fresh air as soon as possible, and artificial respiration given if necessary until a normal breathing rate is restored. 100% oxygen should then be given but not for more than 1 hour.

A medically qualified person or ambulance must be called as soon as possible.

### Skin contact

Irrigate the affected part with a steam of cold tap water for at least 15 minutes. Do not use any antiseptic cream or lotion. Cover with a dry dressing and treat as a thermal burn.

### Eye, nose and throat contact

MEDICALLY QUALIFIED STAFF REQUIRED AS SOON AS POSSIBLE.

Irrigate the eyes at once by placing the head over a sink and directing the stream of cold tap water from a flexible hose onto the face, carefully opening the eyelids while irrigating the eye to remove any trapped ammoniacal liquid. Do not delay – every second is vital to reduce damage to the patient's eyes.

Nose and throat should be irrigated in a similar way for at least 10 minutes. The patient should then be encouraged to drink large volumes of water or lemonade to dilute any ingested ammonia.

---

# Antimony

# Sb

## Stibium

**Exposure limits**
OES = 0.5 mg/m$^3$ air

## Description
Antimony is a silver-white metalloid which is stable at room temperature.

## Properties
Heating antimony metal in air or oxygen produces dense white fumes of antimony trioxide. It has chemical properties similar to those of arsenic, and readily forms alloys with arsenic, tin, iron, zinc, and bismuth.

$AW = 121.76$, $SG = 6.7$, $MP = 630°C$, $BP = 1380°C$.

## Industrial uses
Antimony is added in the preparation of industrial alloys to give extra strength, increased hardness and low coefficient of friction. Metal alloys of this type are used in white metal machine bearings, and in the manufacture of electric cables, lead battery storage plates and ammunition. Very pure antimony, usually obtained from the decomposition of antimony trihydride gas, is now used in the manufacture of semiconductors.

## Detection methods
Contaminated air is passed through a gas scrubber which removes antimony containing vapour into solution which can be analysed by means of atomic absorption spectroscopy.

## Precautions when using the substance
Antimony when used in industry is often associated with the presence of lead either as the metal or one of its compounds. If this is the case, then strict hygiene and worker personal protection as for lead working must be observed (see Lead). Antimony and its compounds are usually present as powders or aerosols and these must be handled carefully to avoid inhalation of dust or vapour during work. Appropriate working overalls with face masks or respirators must be worn unless the operations are carried out inside exhaust protective cabinets.

## Occupational Health
The EOHS reports that antimony usually enters the body through the lungs but can be absorbed through the skin. It is taken up by the blood, particularly the red cells, and tissues and distributed around the body. Within 48 hours of the exposure, the

major part of the antimony is eliminated in the faeces and the urine. The antimony bound to the red cells remains in the body for a much longer time (about 17 weeks) being released when the red cells are naturally replaced. Antimony behaves like arsenic in combining with sulphydryl groups in SH containing enzymes and inhibiting their action in cellular oxidation reactions. The use of antimonial drugs in man has shown disturbances in the normal cardiac rhythm, producing reduction in the T wave and increase in the QT interval.

Exposure to antimony fumes and dusts in miners and smelters can produce inflammation of the mucous membranes, respiratory tract and lungs.

Acute poisoning, which is very rare, produces acute symptoms affecting the gastrointestinal system and producing the abnormal heart rhythm already described, with renal and hepatic complications which can lead to coma and death. Prolonged chronic exposure exhibits a similar but less severe pattern of events often with skin ulcerations.

---

**First aid**

The subject should be removed without delay from the contaminated area and placed on a stretcher, and kept at rest and warm. If the breathing is laboured or difficult, oxygen should be administered intermittently to prevent anoxia. **The patient should be seen by a medically qualified person as soon as possible to determine the degree of follow-up medical treatment.** Acute or significant poisoning by antimony will require the removal of the patient to hospital and the administration of intramuscular injections of dimercaptopropanol (Dimercaprol BAL).

---

# Arsine

# AsH₃

## Arsenic hydride

**Exposure limits**
OES = 0.05 ppm (0.2 mg/m³)
STEL = none quoted
Arsine is one of the most toxic gases known to man. It should be handled with extreme caution.

## Description
Colourless, **unstable, flammable, extremely toxic gas,** heavier than air, with a mild garlic like odour. It is usually formed when nascent hydrogen is liberated in the presence of arsenic.

## Properties
Arsine gas can decompose readily when heated to form an arsenic mirror or a fine deposit of arsenic which is also toxic.

MW = 78, SG = 3.48, MP = –116°C, BP = –55°C, VD = 2.66.

## Detection methods
The thermal decomposition of arsine is used as a simple way of identifying the gas and was used as a simple diagnostic test for the presence of arsenic in forensic samples. Methods now rely on electronic arsine/phosphine monitors, mercury chloride test papers or gas testers using chemical reaction tubes.

## Precautions when using the gas
**The inhalation of arsine gas at concentrations greater than 10 ppm is fatal. The minimum lethal dose for humans is unknown.**

Thirty-two per cent of all arsine poisoning cases have been fatal.

Work involving the use of arsine gas generators or processes that use or may produce arsine must be carried out in specially ventilated laboratories with exhaust protective filter cabinets, which trap any arsine present in the extracted air. No work with arsine should be sanctioned until safe working arrangements have been made and tested.

Arsine is normally supplied in small cylinders which should be stored in a locked, ventilated room. **In an event of a leakage of gas from a cylinder, evacuate the work area and keep all staff away from the contaminated zone.**

**Trained staff using positive pressure breathing apparatus should attempt to shut the cylinder valve or move the cylinder into a filter cabinet or the open air, downwind from staff and buildings. The incident must be reported immediately to the fire brigade.**

## Occupational health

Prevention of exposure to arsine is essential since accidents involving the gas can be severe and fatal.

### Acute exposure

Symptoms of arsine poisoning appear from 20 minutes to 3 hours after the initial exposure. The immediate symptoms are related to the effect on the central nervous system, e.g. giddiness, headache, tingling of extremities, nausea and vomiting. Pain may occur in the chest and abdomen. The urine of the subject, if produced, will be dark and blood stained since arsine is a powerful haemolytic agent.

The patient's skin is coloured with a coppery jaundiced hue with a blue discoloration of the sclera of the eyes. The origins of these symptoms are interesting and related to the biochemical action of arsine.

Arsine, and its related compounds such as bichlorovinyl dichloroarsine or mustard gas, are toxic because they combine rapidly with enzymes containing SH groups to form strong covalently bonded arsenic compounds.

Arsine attacks the red blood cell in two ways:

1. It combines with the SH containing tripeptide, glutathione, which is required to maintain the ion pumping mechanism of the red cell membrane. When this mechanism is affected, sodium ions pass into the red cell causing osmotic swelling and finally rupture of the cell membrane. This haemolysis, which can last up to 96 hours following arsine exposure, accounts for the high plasma haemoglobin levels, leading to the production of the bile pigments bilirubin and biliverdin, seen in jaundice and the blood stained urine.

2. It produces a toxic arsine-haemoglobin complex by reacting with the SH groups on the haemoglobin protein molecule. This complex is non-dialysable and requires the technique of exchange transfusion for its removal.

Arsenic is toxic to many organs particularly the central nervous system, heart, liver and kidneys. The effect on the kidneys is very obvious, producing an arsine-induced anuria (failure to produce a normal urine flow). This anuria may be due to either a direct effect on the tubular cells of the kidneys, or a massive release of an arsenic–haemoglobin–haptoglobin complex which precipitates out in the lumen of the kidney tubule.

---

**First aid**

**ARSINE POISONING IS AN ACUTE MEDICAL EMERGENCY.**

The patient must be removed to hospital by ambulance with minimum delay, preferably with medically qualified staff in attendance, or a person who is able to explain the nature of the poisoning to the hospital emergency staff. The hospital receiving this patient should be alerted as to the nature of the problem.

Administer pure oxygen, give artificial respiration with supplementary oxygen if breathing is weak. Keep the patient warm and at rest.

Modern acute medical treatments consist of exchange transfusions, mannitol diuresis and urinary alkalization and possible peritoneal or haemodialysis if renal failure occurs.

---

# Beryllium  Be

### Exposure limits
OES = 0.002 mg/m$^3$ (This value is under review by the ACTS and WATCH Committees.)

## Description
Beryllium is a light but hard shiny metal resembling steel.

## Properties
It possesses low density and lightness for a metal with high tensile strength and corrosion resistance. It has similar chemical properties to aluminium and magnesium. Finely divided beryllium powder will burn in air.

AW = 9.01,   SG = 1.85,   MP = 1278°C,   BP = 2970°C.

## Industrial uses
Because of its lightness and resistance to corrosion, beryllium is used in the preparation of many specialized alloys particularly beryllium bronzes used in electrical switches and in spark-free impact tools. The metal has been used to moderate the production of thermal neutrons in nuclear reactors, in aerospace construction, and as window material for the manufacture of X-ray tubes. Beryllium phosphors were originally used as a coating for fluorescent electric lamps. Because of the high toxicity of the beryllium powder this practice has now ceased.

## Detection methods
Beryllium in nanogram quantities can be separated by a chemical chelation extraction technique and determined by flameless atomic absorption spectroscopy.

## Precautions when using the substance
As the OES value indicates, **beryllium and its compounds are extremely toxic and very difficult to irradicate from the body once they have entered**. The normal route of entry is via the lungs and not through the gut or undamaged skin. Beryllium containing powder or dust can persist in the lung for periods of up to 20 years. **All beryllium containing powders or its compounds must be handled in a controlled environment.** Small operations can be carried out in a glove box, but larger-scale work requires an exhaust protected cabinet fitted with a suitable dust filter to remove any beryllium from the extracted air. This is designed to protect both the worker and the environment from contamination with beryllium. Where possible, the beryllium should be contained as wet aqueous suspensions thus reducing the health hazard from the dust.

An extremely high standard of cleanness and good housekeeping are essential in laboratories and workshops that handle beryllium. These processes must be segregated from other work, and the staff employed must wear personal protective clothing that is worn only in the workplace and laundered by specialist contractors. Dust masks and/or respirators may be necessary in certain operations but the use of exhaust protective cabinets isolating the work from the staff should mean that these are not normally required.

Fire Precautions
The presence of finely divided beryllium powder is a potential fire risk. Ignition of this powder can be easily caused by heat, friction, and electrical sparks. Fires must be extinguished using dry powder type fire extinguishers. The Fire Service attending a beryllium fire should be informed of the toxic nature of the beryllium powder or its oxide, and firefighters must wear full protective clothing with CABA and be decontaminated under a cold water shower after the fire is extinguished.

## Occupational health
A large amount of occupational health data exists on the effects of exposure by workers to beryllium. This has been compiled from clinical evidence obtained from the mining and refining industry, and the work of occupational physicians working for the atomic energy industry. The EOHS states that the occupational health standard for beryllium is 2 micrograms per cubic metre of air, a level set in 1948 by the US Atomic Energy Commission. Fluctuations in the ceiling value of 5 micrograms per cubic metre of air are permitted as long as the TWA value quoted above is not exceeded. At present, the OES level is being re-examined in the light of the possibility that beryllium may be a human carcinogen. Workers who handle beryllium on a regular basis must undergo pre-employment and periodic medical examinations including chest X-rays to assess the degree of exposure. Beryllium can be detected in the urine of exposed workers by graphite furnace atomic absorption spectroscopy following chelation extraction. The levels reported vary from 10 to 1000 nanograms per litre, and trace amounts of beryllium have been found up to 20 years following exposure.

The toxicology of beryllium is not well understood, but it appears that living tissue interacts with beryllium, which seems to act like an antigen, and stimulates the production of a specific antibody, thus sensitizing the individual. This sensitization is shown as dermatitis and pulmonary hyperreactivity. In sensitized workers, the level of the blood immunoglobulin IgG is found to be significantly raised.

Although beryllium poisoning is now rare in Europe, it is a risk that still occurs in the mining industry. The clinical information concerning beryllium exposure comes from occupational health studies of the manufacture of the phosphor used in the production of fluorescent lamps. In the original production, the fluorescent tubes were coated with a 2% beryllium phosphor. This led to considerable health risks in the manufacture of tubes and also their final disposal. Modern manufacturing processes have replaced beryllium with a calcium halophosphate coating which is virtually non-toxic.

Acute beryllium poisoning produces dermatitis, conjunctivitis, and acute respiratory illness notably dyspnoea, tracheitis, and bronchitis. Removal from the exposure usually leads to remission of clinical symptoms in a few weeks. Bed rest is essential, with supplementary intermittent oxygen, if necessary, until chest X-rays show no abnormalitites.

Chronic beryllium poisoning is more serious. After some months' exposure, signs of weakness, extreme weight loss (11 kg per month) occur. There is extreme dyspnoea with cardiac failure leading to oedema of the arms and legs. About 30% of all cases are fatal, and a further 30% are permanently disabled. There is no satisfactory way of removing beryllium from the body by the use of traditional metal chelating agents.

---

**First aid**

The patient must be removed from the contaminated area without delay by rescuers wearing respiratory and protective overalls. All contaminated clothing must be removed from the victim and the skin washed with cold tap water to remove beryllium dust. If breathing is distressed, oxygen should be given. All cases of beryllium poisoning must be taken to hospital without delay, and seen by a medically qualified person who should contact the nearest Poisons Unit or toxicologist.

---

# Boron trichloride BCl$_3$

**Exposure limits**
OES = none recommended

### Description
Colourless, volatile liquid which fumes in moist air with a choking smell.

### Properties

MW = 117.2,   SG = 1.35,   MP = –107.3°C,   BP = 12.5°C.

This fuming liquid is decomposed by water and ethyl alcohol.

### Detection methods
The presence of the liquid can be detected by the reaction of the hydrolysis product (HCl) with concentrated ammonium hydroxide to produce white fumes. This test is useful to detect leaks on cylinders of gas pipelines.

### Precautions when using the gas
Boron trichloride is an extremely corrosive volatile liquid which is easily hydrolysed to give gaseous hydrochloric acid. Great care should be exercised when using this substance. In laboratories, work with the liquid should be confined to an exhaust protective cabinet. Face and eye protection is essential, nitrile rubber gloves should be worn together with a nitrile apron over a standard laboratory coat.

In industrial plants using boron trichloride, a greater degree of protection is required. Chemical resistant clothing with full respiratory protection is necessary when handling the volatile liquid outside a protective area or pipeline.

### Occupational health
Inhalation of the vapour from boron trichloride will produce oedema and irritation to the upper respiratory tract producing a severe cough.

**First aid**

Remove the patient from the contaminated area, wearing compressed air breathing apparatus and full protective clothing.

## Inhalation

Patients exposed to boron trichloride vapour or its decomposition products will experience severe upper respiratory irritation and coughing. Following removal from the contaminated area, they should be kept warm and at rest, preferably in a sitting position to reduce pulmonary oedema. Breathing can be improved by administration of oxygen or oxygen-enriched air. **The patient should be taken to hospital by ambulance without delay for specialist medical treatment.**

## Skin contact

Physical contact with the liquid or vapour will cause acid burns due to the presence of hydrochloric acid. Immediately remove any contaminated clothing and irrigate the affected area with large volumes of cold tap water from a hosepipe for at least 10 minutes. The patient should then be covered in a warm blanket and any other injuries treated by covering with a dry dressing. Unless the exposure is trivial, the patient should be taken to hospital for further specialist treatment.

## Eye contact
**THIS IS A SERIOUS MEDICAL EMERGENCY.**

The eyes must be irrigated without delay by using large volumes of cold tap water from a hosepipe making sure that the orbits of the eyes are fully washed. Encourage the patient to open and shut the eyes during the irrigation process to enable water to wash away any chemicals from the surrounding tissues. Following irrigation treatment, place a dry dressing or eye pad over the eye(s), **and immediately refer the patient to hospital for specialist medical treatment**.

# Boron trifluoride BF<sub>3</sub>

# Boron trifluoride $BF_3$

**Exposure limits**
OES = none quoted
STEL = 1 ppm (3 mg/m$^3$)

## Description
A colourless gas, heavier than air, with a pungent suffocating odour. **This gas is highly toxic and corrosive.**

## Properties
Boron trifluoride is very soluble in concentrated acids and organic solvents.

MW = 67.8,  SG = 2.99,  MP = –126.7°C,  BP = –99.9°C.

## Detection methods
Hydrolyses in air to form dense white fumes.
Gives dense white fumes in the presence of ammonia vapour.
Infra-red analysis.

## Precautions when using the gas
This gas has similar properties to boron trichloride in that it is easily hydrolysed to give gaseous hydrofluoric acid. **Great care should be used when handling this substance.** In laboratories, work with the gas should be confined to closed apparatus placed inside an exhaust protective cabinet. Face and eye protection are essential, nitrile rubber gloves should be worn together with a nitrile apron over a standard laboratory coat.

In industrial plants, a greater degree of protection is required. Chemical resistant clothing with full respiratory protection is necessary when repair or maintenance work is carried out on pipelines or equipment filled with the gas.

## Occupational health
The principal danger from this gas is from its action on the eyes and mucous membranes. These moist tissues encourage the hydrolysis of the gas and the liberation of hydrofluoric acid which causes irritation and in severe cases destruction of tissue. **Contact with the gas should be avoided at all costs.**

**First aid**

Remove the patient from the contaminated area, the rescuers wearing compressed air breathing apparatus and full protective clothing.

## Inhalation

Persons exposed to boron trifluoride vapour or its decomposition products will experience upper respiratory irritation and coughing. Following removal from the contaminated area into fresh air, they should be kept warm and at rest, preferably in a sitting position to reduce the possibility of pulmonary oedema. Breathing can be improved by administration of oxygen or oxygen-enriched air. **The patient should be taken to hospital by ambulance without delay for specialist medical treatment.**

## Skin contact

Physical contact with the liquid or vapour will cause acid burns due to the presence of hydrofluoric acid. Immediately remove any contaminated clothing and then irrigate the affected area with large volumes of cold tap water from a hosepipe for at least 15 minutes. The patient should then be covered with a warm blanket and other injuries treated by covering with a dry dressing as for a chemical burn. (If calcium gluconate gel is available this should be applied onto the affected skin area prior to covering with a dry dressing.) The patient should be taken to hospital for specialist medical evaluation and treatment.

## Eye contact

**THIS IS A SERIOUS MEDICAL EMERGENCY.**

The eyes must be irrigated without delay by using large volumes of cold tap water from a hosepipe making sure that the orbits of the eyes are fully washed. Encourage the patient to open and shut the eyes during the irrigation process to enable water to wash away any chemicals from the surrounding tissues. If a 10% solution of calcium gluconate is available irrigate the eye(s) with this solution. Following the irrigation treatment, place a dry dressing or eye pad over the eye(s) **and immediately refer the patient to hospital for specialist medical treatment.**

Note: It is a good idea to attach a luggage label to the casualty with the name of the substance involved, and the approximate time the subject was exposed to the gas and the duration of the exposure, if known.

# Bromine

# Br$_2$

**Exposure limits**

OES Long-Term Exposure Limit (8 hour TWA
reference period) = 0.1 ppm (0.7 mg/m$^3$)
Short-Term Exposure Limit (10 minute reference
period) = 0.3 ppm (2.0 mg/m$^3$)

## Description

At room temperature, bromine exists as a dark brown fuming liquid readily producing a reddish suffocating corrosive gas. **Both the liquid and the vapour are extremely hazardous and poisonous. They should be handled with great care.**

## Properties

Bromine is only slightly soluble in water, very soluble in ethyl alcohol, chloroform, diethyl ether, carbon disulphide and carbon tetrachloride. **It reacts with metals and organic tissue and is volatile at room temperature.** Liquid bromine reacts with spontaneous ignition when it comes into contact with potassium, phosphorus and tin.

MW = 159.8, SG = 2.93, MP = −7.2°C, BP = 58.8°C, VD = 5.5,
VP = 175 mmHg (23.27 × 10$^3$ Pa).

## Detection methods

Chemical reaction tubes.

## Precautions when using the gas

Because of the extreme toxic and corrosive nature of bromine liquid and vapour as indicated by the low OES values, all work with this substance must be carried out in a controlled and well-ventilated area. In a laboratory all synthetic chemistry must be carried out in a well-ventilated fume cabinet having an air inlet velocity of 1 metre/ second at the working face. Safety equipment and facilities must be available for controlling spillages of liquid bromine and/or serious leakages from the chemical apparatus used. Personal protective clothing must include chemical resistant clothing with a neoprene apron, gauntlets, eye protection with face visor and chemical resistant boots. In the industrial workplace, the risks of exposure are far greater and therefore a greater degree of personal protection and control of the workplace is required. Where large volumes of bromine liquid are handled, full chemical protection suits fitted with full face visors and positive pressure air lines are required.

## Occupational health

Occupational health and safety data from the EOHS clearly indicate that **bromine vapour can cause acute and chronic poisoning**. Bromine can enter the body via the respiratory system, the skin, and the gastrointestinal tract, acting as a cumulative poison and displacing chlorides and iodides from tissues. The long-term effects of this intoxication include disorders of the nervous system. The low OES set for bromine must be rigorously enforced in the workplace.

A prolonged exposure of $0.5\,mg/m^3$ should not be exceeded (OES = $0.7\,mg/m^3$).
In an atmosphere of 3 to $4\,mg/m^3$, a respirator with a bromine absorbing cannister must be worn.
An atmosphere of 11 to $23\,mg/m^3$ produces severe chocking with no respiratory protection.
An atmosphere of 30 to $60\,mg/m^3$ is extremely dangerous to man without respiratory protection.

Exposure to low concentrations of bromine vapour can result in inflammation of the eyelids, lacrymation, coughing with respiratory distress and headache. These symptoms may be followed by nausea, gastric pain, diarrhoea, and asthma like respiratory attack. Higher concentrations produce more severe effects with inflammation of the tongue and mucous membranes of the upper airways due to chemical attack.

---

**First aid**

Inhalation
Inhalation is the most common form of bromine exposure. The vapour or liquid spillage producing large volumes of bromine vapour produces coughing and chocking in exposed subjects. **Immediately remove the victim from the affected area (rescuers must wear respiratory protection) into fresh air. If breathing stops or is difficult, give artificial respiration with supplementary oxygen using a resuscitator.** Mouth-to-mouth rescuscitation is not recommended. Severe exposure can cause shock, rapid pulse, sweating and collapse. **Keep the patient lying down at rest and warm. Call medically qualified assistance as soon as possible. If not readily available call an ambulance and take the casualty to hospital. Do not leave the casualty until qualified help arrives.**

Skin contact
**Wash the contamination away from the skin and/or clothing with large volumes of cold tap water using a flexible hosepipe or shower unit. Remove all contaminated clothing for the victim.** If not removed quickly bromine liquid can cause deep chemical burns which are difficult to heal and prone to ulceration. **Cover the affected area with a clean dry burns dressing. Arrange for the casualty to go to hospital for medical advice and further treatment.**

Eye contact
THIS IS A MEDICAL EMERGENCY.
 **Irrigate the eyes with large volumes of cold tap water making sure that the eyes are thoroughly washed clean of bromine contamination but taking care to avoid forcing any bromine solution into the eye socket.** Hold the patient's head to one side and allow a steady stream of water from a flexible hosepipe or eyewash station to irrigate the eye(s). **Irrigation should be continued for at least 15 minutes. This is essential if corneal damage is to be reduced. The victim should then be taken to hospital without delay** so that an expert ophthalmalogical opinion can be given on further treatment.

# Butane  $C_4H_{10}$

### Exposure limits
OES = 600 ppm (1430 mg/m$^3$)
STEL = 750 ppm (1780 mg/m$^3$)

## Description
A colourless, highly flammable gas which forms heavier-than-air very explosive gas/air mixtures.

## Properties

MW = 58.1,   VD = 2.0,   BP = −0.5°C,   EL = 1.9–8.5%,   IT = 405°C.

Butane is an easily compressible gas and is supplied in steel cylinders of various sizes containing the liquefied gas. Butane is now used as a propellant gas in many aerosol containers replacing fluorocarbons previously used.

## Detection methods
Flammable gas detector.
Chemical gas detector tubes.

## Precautions when using the gas
**Butane in liquefied form as liquefied petroleum gas is a major potential fire and explosion danger in any factory, workplace or in the home.** Since the explosion limits in air are between 1.9 and 8.5%, and the gas is heavier than air, slight leakages from pipeline connectors or cylinder valves can easily form an explosive mixture and give rise to explosions and fires. It should not be used in unventilated basements or rooms. **(Note: 500 grams of the gas mixed with air has an explosive equivalent of about 1 kilogram of TNT.)**

**Leakage of a standard size cylinder of liquid gas (25 kg) into a workshop or laboratory, allowed to mix with air and ignited, would destroy the building and kill any occupants.**

Butane cylinders should be stored in the open air, protected from heat and sunlight and the gas supply should be piped into the building, if possible, by fixed gas lines. Only small size gas cylinders, e.g. up to 6 kg, should be used inside buildings as a portable gas supply. **All compression joints and union couplings to the cylinder and equipment must be soap solution tested for leaks, daily and every time the fittings are disturbed.**

## Occupational health

The OES and STEL values indicate that short exposures to butane (10 minutes) of concentrations equivalent to 10 000 ppm (1%) produce slight drowsiness, but no other adverse effects. Higher concentrations will cause a dry irritation in the throat, tightness of the chest, and in severe cases, mental confusion leading to unconsciousness, probably due to the reduced oxygen concentration in the contaminated air.

---

**First aid**

Inhalation
The subject should be removed from the contaminated area into fresh air and allowed to recover. If the subject is drowsy or unconscious, oxygen-supplemented artificial respiration should be given until breathing is normal and regular and the patient appears well oxygenated as shown by pink nailbeds or lips.

Skin contact
Liquid butane volatilizes readily from the skin but may cause a mild form of cold burn. If the spillage has saturated clothing, this should be removed from the subject, and the affected area irrigated with a stream of cold tap water for at least 10 minutes. The skin should be examined and if there is irritation and redness at the affected site with blisters, the casualty should have a dry burns dressing applied to the injury and be referred to the nearest Accident and Emergency department.

Eye contact
**Splashes of liquefied gas into the eye is a serious matter.** Immediately treat the casualty by irrigating the eye with a stream of cold tap water to prevent further tissue damage. The irrigation should be continued for at least 10 minutes. Cover the eye with a dry eye pad, do not place any medication in the form of drops into the eye, **and take the patient to hospital without delay.** Inform the medical staff at the hospital of the nature of the exposure and the substance involved.

---

# Cadmium Cd

**Exposure limits**

OES cadmium compounds and cadmium oxide
fume = 0.05 mg/m³
cadmium sulphide and respirable dust =
0.04 mg/m³

### Description
A soft silver-white metal with a blueish tinge.

### Properties
The metal is malleable and ductile, soluble in acids but insoluble in water.

AW = 112.4,   SG = 8.64,   MP = 320.9°C,   BP = 765°C.

Burns in air or oxygen to produce a brownish yellow oxide.

### Industrial uses
Because it is highly resistant to corrosion, cadmium is used to coat or electroplate metals such as iron and steel. Some 50% of all cadmium is used for this purpose. Large amounts are used for colouring in paint and plastic manufacture, in the production of resins, and as a component in the nickel-cadmium alkaline battery.

### Detection methods
Colorimetric analysis using di-beta naphthylthiocarbazone following sulphuric–nitric acid digestion.

Polarographic analysis of nitric–perchloric acid sample digest has the advantage of simultaneous determination of lead and cadmium in the same sample.

Cadmium levels in blood and tissues are now usually measured by graphite furnace atomic absorption spectroscopy following acid digestion of the samples.

### Precautions when using the substance
The values for the OES of cadmium compounds indicates the extreme toxicity of cadmium. Medical evidence also shows that any cadmium taken up by the body is very slowly eliminated. The biological half life has been estimated as between 7 and 30 years. This means that work with cadmium or its salts must be very carefully regulated particularly ensuring that cadmium is not inhaled, ingested or absorbed through the skin during work operations. Precautions similar to those required under the Lead Regulations should be imposed. No worker should be exposed to cadmium fumes without adequate dust control and exhaust ventilation. The exhaust ventilation system used must be fitted with an extraction filter(s). In work areas with the possibility of high contamination risk, e.g. the changing of air extraction filters, the wearing of special dust respirators will be necessary. Good personal hygiene

must be encouraged in all cadmium workers by providing good washing and clothes changing facilities. Work overalls and contaminated clothing must be placed in special bags and laundered to remove cadmium dust. Staff must be instructed to wash and change their clothing before meals, and facilities for a hot shower should be available at the end of the work shift.

## Occupational health

The extreme toxicity of cadmium is due to its action as an inhibitor of enzyme systems in the body that possess SH groups principally represented by the cellular dehydrogenase system. This allows normal cellular oxidation to occur which is vital to the life of cells. EOHS suggest that under normal conditions, 20–50% of any cadmium dust inhaled can be absorbed through the lungs. Absorption via the gastrointestinal system is much less, between 2 and 5% of the amount ingested. In iron deficiency, the amount absorbed can increase to 20%. Any cadmium absorbed in the body is transported to the liver and formed into a low molecular weight protein complex, cadmium metallothionein, which is thought to protect the body from the toxic effects of free cadmium ions. Cadmium concentrates in the kidneys and the liver which together contain about 50% of the cadmium present in the body.

## Acute poisoning

EOHS states that inhalation of air contaminated with cadmium above $1\,mg\,Cd/m^3$ for 8 hours or higher concentration for shorter periods of exposure, may lead to chemical pneumonitis and pulmonary oedema. Symptoms develop within 8 hours from exposure and have the chacteristics of fume fever. Death may occur up to 7 days from the initial acute exposure. Levels of exposure exceeding $5\,mg\,Cd/m^3$ may occur in the smelting, welding or soldering of cadmium. In 1966, all five steel erectors cutting through cadmium coated steel bolts on the Severn Road Bridge were acutely poisoned by cadmium fumes from the operation. They were working in an enclosed space for about 5 hours.

## Chronic poisoning

Prolonged exposure to cadmium dust, fumes and cadmium compounds produces chronic cadmium poisoning. Occupational health studies show that heavy exposure to cadmium in the air (air concentrations $>0.1\,mg\,Cd/m^3$) results in loss of weight, cough and dyspnoea, and gross pulmonary emphysema. Lower levels of exposure over a long period show profound kidney dysfunction and damage because the kidneys is the organ first affected as the cadmium accumulates in the renal cortex. Levels of cadmium $>100\,\mu g\,Cd/gram$ wet weight cause tubular cell damage, with an inability for the renal tubules to reabsorb protein. Cadmium is lost from the kidney tubules associated with a low molecular weight protein known as $B_2$ microglobulin.

Excessive chronic exposure to cadmium in contaminated food, particularly rice, produces a painful form of osteomalacia causing multiple bone fractures together with serious kidney dysfunction. This disease, known as itai-itai, has been reported in Japan and is caused by eating food grown in fields having high cadmium contamination because of their proximity to cadmium smelting works.

## Health screening for cadmium workers

All persons handling or processing cadmium must be regularly monitored for cadmium exposure. Blood cadmium levels indicate exposure during the last few months. Blood cadmium values above $10\,ng\,Cd/ml$ of whole blood ($1\,\mu g\,Cd/100\,ml$) must be considered clinically significant if the period of exposure is long. The level in urine is used to estimate the body burden for cadmium. This value should not exceed $10\,\mu g\,Cd/g$ urinary creatinine (about $10$–$20\,\mu g\,Cd$ per day). Medical examinations should be carried annually for all workers exposed to cadmium. In screening for exposure, the level of $B_2$ microglobulin should be determined in addition to the blood and urine estimations. This value should be less than $0.5\,mg/l$.

---

**First aid**

Persons exposed to cadmium must be removed from the contaminated area without delay, the rescuers wearing appropriate protective clothing and breathing apparatus (CABA type).

Inhalation of fumes or vapour

The victim should be placed on a stretcher at rest and if the inhalation is acute should receive oxygen-enriched air of pure oxygen from a resuscitator as required. If pulmonary oedema is seen or suspected, continue positive pressure oxygen treatment and call for medically qualified assistance. **The patient should be removed to hospital without delay.**

Ingestion of cadmium salts

Conscious patients should be encouraged to vomit to remove any cadmium from the stomach. The possibility of gastrointestinal irritation can be reduced by getting the patient to drink milk and/or beaten eggs. **The victim should be removed to hospital without delay for further symptomatic treatment and expert toxicological advice.** There is at present no specific treatment for cadmium poisoning. The use of chelating agents is contraindicated because the cadmium chelates formed are toxic to the kidneys.

---

# Carbon dioxide $CO_2$

### Exposure limits
OES = 5000 ppm (9000 mg/m³)
STEL = 15 000 ppm (27 000 mg/m³)

## Description
A colourless, odourless gas with a soda water taste. It is present in normal air in concentrations ranging from 0.03 to 0.06% by volume.

## Properties
Carbon dioxide dissolves in water to produce carbonic acid which is the basis of the carbonated soft drink industry. Carbon dioxide gas can be compressed to produce liquid carbon dioxide and rapid cooling of the liquid form easily produces solid carbon dioxide. In medicine, liquid carbon dioxide is used as a simple topically applied freezing solution for minor surgery and in the operation of the freezing microtome used to produce frozen histological sections for diagnostic purposes. Solid carbon dioxide is used as a refrigerant in the manufacturing industry and in the transport of goods and chemicals which are unstable or decompose at room temperatures.

Carbon dioxide gas is effective in removing oxygen and cooling small fires and is used in portable fire extinguishers for this purpose.

MW = 44, SG = 1.98, MP = −56.6°C at 5.2 atmos., BP = −78.5°C (sublimes), VD = 1.53.

## Detection methods
Chemical gas detector tubes.
Infra-red gas analyser.

## Precautions when using the gas
Carbon dioxide is 1.5 times heavier than air and easily reduces the available oxygen content of the air in confined spaces such as basements and sewers. This effect is insidious as the gas cannot be detected by smell and the first signs of carbon dioxide affecting the subject is their collapse due to anoxia. No work which can liberate carbon dioxide gas should be carried out in poorly ventilated areas. Liquid and solid carbon dioxide are cryogenic substances and must be handled with the appropriate safety gloves and eye protection. Failure to do so can give rise to painful cold injuries.

## Occupational health

At concentrations below the OES, no adverse effects are seen. Concentration of 5% (50 000 ppm) can produce headaches and shortness of breath. At concentrations greater than 10% carbon dioxide, unconsciousness will occur due to oxygen deficiency. Research by the Navy shows that exposures of submariners, who may breathe levels of gas up to 1% carbon dioxide by volume for long periods, can produce calcium deposition in body tissues, e.g. the kidneys, possibly due to mobilization of calcium normally present in the bones. In nuclear submarines, great care is taken to prevent any build-up of the carbon dioxide present in the air due to the long duration of voyages.

---

**First aid**

Inhalation

**Remove the casualty from the contaminated area remembering that the rescuers must wear self-contained compressed air breathing apparatus (under no circumstances should gas masks be worn). Remember – rescuers' lives have been lost by ignoring this instruction.** If the victim is unconscious, it is almost certain that this is due to oxygen deficiency. Immediately give the patient oxygen either as 100% oxygen by means of a face mask or by an oxygen-enriched resuscitator. Remain with the casualty until breathing has returned to a normal depth and rate. The casualty should be seen by a medically qualified person to ensure a full recovery has been made.

Skin and eye contact with liquid or solid carbon dioxide

The greatest danger is from their cryogenic action on tissues. It is important that the affected area is treated by drenching with a steady stream of cold tap water maintaining this treatment for at least 15 minutes. If the eye(s) are involved, or the skin area affected is not trivial in extent, all casualties should be referred to hospital without delay to receive further expert care.

---

# Carbon disulphide $CS_2$

**Exposure limits**
OES = 10 ppm (30 mg/m³)
STEL = none

## Description
A colourless, poisonous, extremely flammable and explosive gas, usually produced from liquid carbon disulphide. Possesses a very unpleasant characteristic smell. It is a highly reactive substance.

## Properties
Carbon disulphide liquid is slightly soluble in water, soluble in ethyl alcohol and diethyl ether.

MW = 76, SG = 1.26, MP = −110.8°C,
BP = 46.3°C, VD= 2.6, VP = 360 mmHg (46.8 × 10³ Pa) at 25°C,
FP = −30°C, EL = 1.3–50%, IT = 100°C.

## Detection methods
Chemical reaction tubes.
Infra-red analysis.

## Precautions when using the gas
Although the gas is toxic, the greatest danger in its use is the risk of fire and explosion even at room temperature. A hot glass rod can ignite carbon disulphide vapour in air. Only small quantities of carbon disulphide liquid should be stored, unless a specially prepared and ventilated area is available. Extra fire fighting equipment must be provided to deal with any emergency.

Carbon disuphide vapour can be easily absorbed in the lungs and about 30% of the inhaled material is retained when a steady state of inhalation is reached. The liquid is easily absorbed through the skin and great care should be taken to wear adequate chemical protective clothing particularly gloves when handling the liquid.

## Occupational health
The EOHS gives very detailed information on the effects of industrial exposure to carbon disulphide. It is primarily a neurotoxic poison and therefore any exposure to the liquid or vapour can result in symptoms indicting central or peripheral nerve damage.

Acute exposures (>15 mg/litre, 4800 ppm) are lethal after 30 minutes.

Sub acute exposures (>2 mg/litre, 640 ppm) seem to produce manic depressive mental disorders possibly with polyneuritis.

Chronic exposures to carbon disulphide at a level of 0.3–0.5 mg/litre over a period of several years leads to weakness of lower limbs, fatigue, headache, sleep disturbances, and parathesia. In male workers there can be hypospermia and decreased excretion of 17 keto, 17 hydroxycorticosteroids and androsterone. Female workers show menstrual disturbances and increased frequency of abortion possibly related to the carbon disulphide easily passing through the placental barrier to the foetus.

Carbon disuphide is readily metabolized in the body to form dithiocarbamates which chelate with divalent metals such as copper and zinc present in blood protein complexes leading to increased excretion of metal-dithiocarbamates in the urine.

---

**First aid**

Inhalation

The casualty should be removed from the source of contamination by rescuers wearing respiratory protective equipment, into fresh air. The patient should be kept comfortably warm and given oxygen-enhanced resuscitation if required. Unless the exposure has been trivial, the casualty should be removed to hospital for a complete medical examination to ascertain the degree of possible neurological damage.

Skin contact

Any skin contact with carbon disulphide should be treated with irrigation of the affected area with a stream of cold tap water for at least 15 minutes. Any contaminated clothing must be removed to prevent inhalation of the toxic vapour by victim and rescuers. The affected skin area should be covered with a dry dressing and the victim referred to hospital for further treatment.

Eye contact

**Eye contact with carbon disulphide liquid or vapour is a serious matter.** Immediately irrigate the eyes with a steady stream of cold tap water for at least 15 minutes to remove as much of the carbon disulphide as possible. **Refer the victim, without delay to the nearest accident and emergency hospital for specialist medical treatment.**

---

# Carbon monoxide CO

## Exposure limits
OES = 50 ppm (55 mg/m$^3$)
STEL = 300 ppm (330 mg/m$^3$)
Author's note: These OES values have been challenged by the American Government and other agencies. An OES of 35 ppm with a OES(C) of 200 ppm seems more appropriate in view of the chronic long-term effects of the gas.

## Description
Colourless, odourless, toxic and flammable gas.

## Properties
The gas forms very toxic metal carbonyls particularly with iron and nickel. Carbon monoxide is produced by many industrial processes and hydrocarbon fuelled engines.

MW = 28.01,   SG = 1.25,   MP = −205.1°C,   BP = −191.5°C,
EL = 12.5–74%,   IT = 608.9°C.

## Detection methods
Chemical reaction tubes.
Carbon monoxide electronic fuel cell detector.
Infra-red analyser.

## Precautions when using the gas
Carbon monoxide poisoning is a significant risk to workers in a number of industries. Carefully pre-planned safety measures are required. Correct type and well-maintained breathing apparatus, available near to the risk area is essential, e.g. a type N gas mask is acceptable provided that the risk area atmosphere will never have less than 18% oxygen by volume. A preferred safety measure would be to have a locally trained rescue team with compressed air breathing apparatus.

## Occupational health
### Acute effects
Carbon monoxide behaves as chemical asphyxiant. It has 300 times the affinity of oxygen for combining with the red cell oxygen transport pigment, haemoglobin. Carbon monoxide rapidly forms a cherry red carboxyhaemoglobin complex and its presence limits the oxygen carrying capacity of the blood. The following table indicates the clinical effects produced by increasing levels of carbon monoxide in the air:

59

| ppm CO in air | % carboxy-haemoglobin in blood | Clinical effects |
|---|---|---|
| 100 | 0 to 10 | No symptoms. Can be tolerated for 8 hours. |
| 500 | 30 to 40 | Headaches, nausea, irritability, increased respiration and weakness after 1 hour. |
| >1000 | 60 to 80 | Loss of consciousness, respiratory failure and death. |

## Chronic effects

**Medical evidence suggests that individuals with coronary heart disease should not be exposed to carbon monoxide concentrations of greater than 35 ppm (i.e. 5% carboxyhaemoglobin).** This gas concentration may initiate or enhance serious degradatory changes in the subject's heart muscle, due to impairment of the oxygen supply to the heart muscle from circulating oxyhaemoglobin. Heavy cigarette smokers often have a blood carboxyhaemoglobin concentration of 5% prior to any further carbon monoxide exposure.

---

**First aid**

**Remove victim from the contaminated area using approved breathing apparatus.** If the patient is breathing, administer pure oxygen using an auronasal mask. If the subject's breathing is shallow or stopped, use an oxygen rescuscitator or give artificial respiration with intermittent oxygen.

Keep the patient warm and at rest. Call an ambulance and remove subject to hospital. Most cases who survive require 48 hours observation and bed rest.

---

# Carbonyl fluoride      COF$_2$

### Exposure limits
OES = 2 ppm (5 mg/m$^3$)
STEL = 5 ppm (15 mg/m$^3$)

## Description
Colourless gas, heavier than air, with a pungent odour. **It is a highly toxic and corrosive gas as hydrolytic decomposition produces hydrofluoric acid.**

## Properties
Very similar to those of hydrogen fluoride, since it is easily decomposed to give hydrofluoric acid in contact with water.

MW = 66,   MP = −83.1°C,   SG = 1.139.

## Detection methods
Wearing positive pressure breathing apparatus, spray suspected area with concentrated ammonia solution. Dense white fumes indicate the presence of hydrofluoric acid.

## Precautions when using the gas
The gas must be treated in the same way as handling hydrofluoric acid. **Great care is required in the storage, handling and use particularly when large quantities are involved.** All staff must be instructed on the hazardous nature of the substance and fully trained in the safety measures and personal protective equipment needed.* Cylinders of the gas should be placed in the open air in a secure metal cage, or in a mechanically ventilated room or exhaust protective cabinet. Connections to apparatus or pipelines should contain an anti-suck-back device. Commercial suppliers of this gas recommend that all users have positive pressure compressed air breathing apparatus available in case of an emergency.

## Occupational health
**A vapour concentration of 50 ppm in air breathed for more than 30 minutes is fatal.** Lower concentrations cause irritation to the mucous membranes and respiratory tract. The eyes are particularly sensitive to carbonyl fluoride vapour. **Inhalation of**

---

*The level of personal protection needed will depend on the scale of operations being carried out. For small laboratory work, this would require a neoprene apron, gloves and boots worn over a chemical resistant overall, with face and eye protection. For large industrial operations, a fully enclosed chemical resistant suit with airline hood would be suitable. Staff handling this sort of dangerous material must not work alone. In factories, it is essential to provide adequate fast acting safety showers, drenches and eye wash stations.

low concentrations of the gas may not appear to have any effect but lung oedema may occur 12–24 hours following the exposure.

---

**First aid**

Exposure to carbonyl fluoride should be considered a serious medical emergency. The presence and/or advice from a medically qualified person is required as soon as possible.

Inhalation

The victim must be removed from the contaminated area without delay, the rescuers wearing protective clothing and compressed air breathing apparatus. Even with minor exposures, the casualty should be given 100% oxygen by means of a oxygen resuscitator at intervals of up to 4 hours following exposure. **In severe exposure, the victim should receive oxygen therapy immediately, and then removed to hospital by ambulance without delay.** A member of staff should accompany the patient to hospital to give details to the doctor of the type and duration of the exposure to assist in the decision on further treatment.

Skin contact

**Immediately drench the affected area or the whole body with a shower of cold water. Remove the victims clothing while under the water shower. Continue cold water drenching for at least 10 minutes to remove as much fluoride from the skin as possible. Get an assistant to seek medical help as soon as possible.**

Treat the affected area with iced 70% ethyl alcohol (ethanol) or an iced saturated solution of magnesium sulphate for 30 minutes. **If medical aid is not available, continue the iced alcohol or magnesium sulphate treatment for 2 hours, and then apply a paste of magnesium oxide in glycerine. Alternatively, use a liberal application of 10% calcium gluconate gel to the affected area.**

Medical note: Some medical practitioners prefer to give subcutaneous injections of 10% solution of calcium gluconate but the use of calcium gluconate gel has proved to be effective in treating liquid splashes and vapour burns.

Eye contact with the vapour or liquid

**This is a medical emergency. Irrigate the eyes and face without delay with large volumes of cold tap water for at least 15 minutes. Get an assistant to call an ambulance. Repeat the irrigation treatment if necessary. The subject must then be removed to hospital and seen by an ophthalmic specialist.**

---

# Carbonyl sulphide COS

### Exposure limits
OES = not quoted (US Government Hygienists
give a value of 10 ppm)

## Description
A colourless, acid flammable and toxic gas.

## Properties
The gas decomposes and burns with a bluish flame when exposed to moisture or alkalis, forming carbon dioxide and hydrogen sulphide, and can form explosive mixtures with air.

MW = 60.07,  SG = 1.07,  MP = −138.2°C,
BP = −50.2°C,  VD = 2.1,  EL = 12 to 29%.

Soluble in ethanol and toluene, but very soluble in carbon disulphide.

## Detection methods
No specific methods are described.

## Precautions when using the gas
Cylinders of the gas should be placed either in a secure metal cage in the open air or in a mechanically ventilated exhaust protective cabinet. Connections to apparatus or pipelines must be fitted with an anti-suck-back device. **Suppliers of this gas recommend the availability of positive pressure compressed air breathing apparatus for use in emergencies.**

## Occupational health
**Carbonyl sulphide is known for its high irritancy and toxicity.** If inhaled, the substance acts on the central nervous system, possibly the respiratory centre in the brain causing respiratory paralysis and death. Carbonyl sulphide decomposes in the lungs producing hydrogen sulphide which is readily absorbed into the blood. All oxysulphides are powerful lung irritants and produce pulmonary oedema in moderate and high concentrations.

**First aid**

Inhalation

Remove victim from the affected area to fresh air, the rescuers wearing compressed air breathing apparatus. **Do not leave the casualty unattended.** If the patient's breathing becomes weak or stops, give artificial respiration preferably with an oxygen-supplemented resuscitator. Keep the victim warm with a light blanket to reduce the possibility of shock. **The casualty must be taken to hospital as soon as possible following the exposure.**

Skin contact

Treat immediately by irrigating the affected part with a stream of cold tap water preferably from a hosepipe for at least 15 minutes, removing all contaminated clothing during the washing process. Cover any skin burns with dry dressings, keep the victim warm, **and transfer to hospital without delay for further treatment from a medical specialist**.

Eye contact

Treat any affected eyes with a stream of cold tap water for at least 15 minutes, ensuring that the irrigation process removes all the contaminating chemical from the surface of the eyes and the surrounding tissues. Cover the eyes with a dry dressing **and remove the victim to hospital without delay for specialist medical treatment**.

# Chlorine

# Cl$_2$

**Exposure limits**
OES (8 hour TWA) = 0.5 ppm (1.5 mg/m$^3$)
STEL = 1 ppm (3 mg/m$^3$)

## Description
Liquid chlorine is an amber coloured liquid about 1.5 times as dense as water. **Gaseous chlorine is a greenish yellow toxic gas, non-flammable but acting as a strong oxidizing agent.** Liquid and gaseous chlorine are used extensively in industry often in very large amounts.

## Properties
Chlorine gas is slightly soluble in water, soluble in alkalis and has a disagreeable suffocating odour at concentrations >3.5 ppm in air.

MW = 70.91, SG = 1.56 (liquid at −34.6°C),
MP = −101°C, BP = −34.6°C, VD = 2.5, VP = 4800 mmHg (638.4 × 10$^3$ Pa), at 20°C.

## Detection methods
Leaks from gas or liquid containing cylinders can be detected using concentrated ammonia solution which produces dense white fumes.
Chemical reaction tubes.

## Precautions when using the gas
**Because of the dangers that can arise to staff and members of the public if a serious spillage of chlorine occurs, all staff must be trained in the use of the chlorine containing equipment or plant, and in emergency repair and safety equipment.** In any operation where significant amounts of chlorine could be released in an accident, provision must be made for easy access to compressed air breathing sets or approved respiratory protective filter masks and oxygen resuscitation apparatus.

## Occupational health
Chlorine gas is a severe irritant to all biological tissues, particularly epithelia covering the lungs, nose, throat, and eyes. Chlorine containing vapour is so irritant that concentrations in air greater than 3 ppm can be readily detected by smell, bearing in mind that workers in chlorine containing atmospheres can develop olfactory fatigue.

Acute exposure gives rise to irritation of the mucous membranes in the eyes, nose and throat leading to burning pains in the chest, leading to reflex coughing and, if severe, vomiting. Fatal cases of acute poisoning reported have been due to massive leakages of gas from large cylinders. This exposure would rapidly lead to pulmonary oedema, fall in blood pressure and cardiac arrest. Chlorine gas concentrations in excess of 40 ppm in air can be highly dangerous if inhaled for more than 30 minutes. Levels of 1000 ppm or greater are lethal even after a few breaths. Levels of gas below 15 ppm in air will produce irritation to the throat with coughing but if the subject is removed from the contaminated environment will not produce any serious lasting effects.

---

**First aid**
**Remove the subject from the contaminated area into fresh air, the rescuer preferably wearing compressed air breathing apparatus.**

Inhalation
Administer pure oxygen to the subject. If breathing is weak or stopped, give artificial respiration preferably using an oxygen-supplemented resuscitator. Keep the casualty at rest, quiet and warm, if possible lying on a stretcher. **Remove the casualty to hospital or call medically qualified staff as soon as possible.**

Skin contact
Chlorine liquid or gas will produce serious chemical burns in contact with skin. Immediately, drench the affected area with a steady stream of cold tap water for at least 15 minutes to remove all the contaminating chemical. If the skin is undamaged, cover with a dry dressing, if there is a serious chemical burn, place a burns dressing or cover with a sheet of clingfilm **and immediately refer the victim to hospital for specialist medical treatment**.

Eye contact
**Irrigate immediately with large volumes of cold running tap water ensuring the globe and recesses of the eye are thoroughly irrigated (holding eyelids back by force if necessary). Spillage into or contact with the eye is a medical emergency and the casualty should be seen by a medical eye specialist without delay.**

Author's note: Workers using or producing chlorine should be aware that in many reactions involving the generation of chlorine gas in the presence of acids containing oxygen, e.g. sulphuric or perchloric acids, the very toxic gas chlorine dioxide may also be evolved. The OES for chlorine dioxide is 0.1 ppm $(0.3 \, mg/m^3)$.

---

# Chromium and compounds Cr

## Exposure limits

| | |
|---|---|
| For chromium metal and insoluble salts | OES (TWA) = 1 mg/m$^3$ |
| For chromic oxide | OES (TWA) = 0.5 mg/m$^3$ |
| For sodium and potassium chromate | OES (TWA) = 0.025 mg/m$^3$ |
| For chromium trioxide | OES (TWA) = 0.025 mg/m$^3$ |
| For lead chromate | OES (TWA) = 0.001 mg/m$^3$ |

## Description

The chemistry of chromium is complicated because the element exists in three valency states, namely, chromous (Cr2$^+$) unstable and easily oxidized to the stable chromic (Cr3$^+$) compounds. These and chromic acid and the dichromates where chromium exists as Cr6$^+$, form the majority of chromium compounds used in industrial plants and laboratories. Most chromium compounds, particularly the dichromates, are strongly coloured.

## Properties

**Chromium (Cr)** is a very hard lustrous steel grey metal which is very resistant to oxidative corrosion. The metal is insoluble in water but soluble in dilute sulphuric and hydrochloric acids.

AW = 52,   SG = 7.2,   MP = 1890°C,   BP = 2482°C.

**Chromic oxide ($Cr_2O_3$),** a very stable green powder used in pigments. Insoluble in water, alkalis, and ethyl alcohol.

MW = 152,   SG = 5.21,   MP = 2435°C,   BP = 4000°C.

**Sodium chromate ($Na_2Cr_2O_7$, $2H_2O$),** a bright orange crystalline material, very soluble in water, insoluble in ethyl alcohol. Sodium chromate is the material from which many chromium compounds are produced.

MW = 298,   SG = 2.52,   MP = 256.7°C in anhydrous form,   BP = 400°C with decomposition.

**Potassium dichromate ($K_2Cr_2O_7$),** a stable orange red crystalline substance, very soluble in water, insoluble in ethyl alcohol.

MW = 294.2,   SG = 2.67,   MP = 398°C.

**Lead chromate ($PbCrO_4$),** an orange yellow pigment known as chrome yellow. **Lead**

chromate is known to be an industrial carcinogen and therefore must be handled with great care. For this reason, the OES is extremely low.

MW = 323.2, SG = 6.3, MP = 844°C, BP = decomposes.

**Chromium trioxide ($CrO_3$) also called chromic anhydride or chromic acid.** The trioxide is formed by the reaction of a concentrated solution of a dichromate with excess sulphuric acid. The chromic acid solution formed has been used as a glass cleaning agent and as a constituent in chromium plating solutions. **It is a violent oxidizing agent.**

MW = 99.99, SG = 2.7, MP = 196°C, BP = decomposes.

## Industrial uses
Chromium metal is used in the manufacture of stainless steel and other corrosion resistant alloys. The metal has been extensively used in the chromium plating of steel particularly in the motor vehicle and the electrical manufacturing industries. Chromium compounds are used in the dyeing and tanning trades and in the preparation of pigments and paints.

## Detection methods
Colorimetric analysis using diphenyl carbazide as a complexing agent for chromium ($6^+$), oxinate, or methyloxinate or acetyl acetone for chromium ($3^+$) compounds.
Flame emission spectroscopy.
DC arc spectroscopy.
Atomic absorption spectroscopy following wet digestion and chelation of the sample with ammonium pyrollidine dithiocarbamate, particularly if it contains a biological matrix.

## Precautions when using the substance
Any processes or operations involving chromium or its compounds require strict dust control measures to prevent inhalation, and precautions to eliminate skin contact with solid or liquid chromium compounds. Dust respirators must be approved for chromium work and capable of retaining 90% of 0.5 micron dust particles. All clothing must be laundered daily by the employer, and all safety equipment inspected and checked for damage each day before a working shift.

## Occupational health
Occupational hazards arise as a result of inhalation of chromium containing dusts, fumes or mists. Studies show that chromium ($3^+$) salts are less toxic than those of chromium ($6^+$). Chromium ($6^+$) compounds are very irritant and corrosive particularly as chromic acid. This form of chromium can be absorbed through the skin, lungs, or the gastrointestinal tract. Chromium ($6^+$) fumes can be formed during the welding of stainless steel.

### Chronic exposure
The commonest condition resulting from exposure to chromium is dermatitis and chrome ulceration. Chromium ($6^+$) salts are known to give rise to skin irritation in the form of a chemically induced dermatitis, and in some cases to sensitization of the subject to chromium. This can be easily established by patch testing the arm of the subject with a 0.5% sodium dichromate solution. Dermatitis is usually seen around the wrists and neck of workers who handle chromium. Remember: a large

number of industrial lubricants and oils have chromium in their formulation, and these products can give rise to irritation leading to contact dermatitis.

Chrome ulceration is more serious. EOHS suggests that this arises from the entry of chromium ($6^+$) through cuts in the skin which then damages it producing deep circular ulcers which may penetrate to the underlying bone if untreated.

---

**First aid**

Inhalation

Inhalation of dust or fumes produces irritation of the mucous membranes with shortness of breath and wheezing. The victim should be removed from the contaminated area into fresh air, placed onto a stretcher, with the patient's back raised to assist breathing and to minimize pulmonary oedema, and oxygen or oxygen-supplemented air given, at intervals, to assist breathing. **The victim should be removed to hospital by ambulance as soon as possible, informing the ambulance paramedical staff of the nature of the exposure.**

Ingestion

Accidental ingestion of chromium salts usually produces nausea, vomiting and acute abdominal pain. Significant amounts of chromium compounds swallowed have a corrosive action on the lining of the stomach. The victim, if fully conscious, should be encouraged to drink half a litre of milk so that the milk protein casein present can buffer the effects of the acidic chromium. Vomiting should be encouraged to remove as much chromium from the stomach as possible and reduce its absorption into the body. Severe poisoning can cause kidney and liver damage. **The casualty must be removed to hospital without delay for further specialist medical treatment.**

Skin contact

Concentrated chromium solutions can produce painful burns if left in contact with the skin. Remove the casualty to a safe area, remove any contaminated clothing, the first aider ensuring that contact with any chromium containing liquid is avoided by wearing vinyl gloves. Chrome ulcers can be prevented by ensuring that any chromium salts are washed from the surface of the skin without delay. Contaminated skin, especially if there are abrasions and/or cuts must be treated at once by copious irrigation of the affected area with cold tap water for at least 15 minutes, followed by the application of 10% calcium disodium versenate ointment. (The versenate ointment converts any chromium ($6^+$) to the ($3^+$) form which is readily chelated by the excess versenate forming a stable water soluble chelate with any chromium present in the tissues.) This treatment will ensure rapid healing of the skin. The casualty should be referred to the nearest hospital for further specialist treatment, ensuring that the receiving doctor is aware on the nature of the chemical involved.

Eye contact

**Any eye contact with chromium solutions, especially chromic acid, must be regarded as a medical emergency.** Immediately irrigate the eyes and surrounding tissues with a steady stream of cold tap water for at least 15 minutes. At the end of the treatment, if the eyes are still painful, or the chromium salts have not been removed, repeat the water irrigation process. Apply a dry eye dressing to protect the eyes, do not place any drops or lotion into the eyes, and take the casualty to the nearest hospital for further treatment by an eye specialist.

---

# Copper and salts Cu

### Exposure limits
Copper (as fumes)    OES (TWA) = 0.2 mg/m³
                     No STEL quoted
Copper (as dusts or mists)    OES = 1 mg/m³
                              STEL = 2 mg/m³

## Description
Copper exists as a reddish brown metal, very stable, resisting oxidation in air, with many uses in industry as an alloy with non ferrous metals, and as copper salts.

## Properties
Copper metal is malleable, ductile and corrosion resistant. It conducts electricity extremely well and is used in the manufacture of electrical contacts and cables. It easily forms alloys with many metals, notably with zinc (brass), tin (bronze), nickel (monel metal), all of which have industrial importance.

AW = 63.5,   SG = 8.92,   MP = 1083°C,   BP = 2567°C.

## Industrial uses
The electrical industry uses about 75% of the world's production of copper. Copper is produced from copper containing ores by smelting at 1500°C and treated to remove sulphur and iron impurities. The resulting blister copper is 95% pure, but is further purified to 99.9% by electrolysis for electrical use.

Many compounds of copper are used in industry. Copper sulphate is used as a fungicide and as part of the formulation in paint as an anti-fouling agent removing barnacles from ship's bottoms. Copper sulphate neutralized with lime is used to spray on vines to prevent mildew (Bordeaux mixture). Copper sulphate is extensively used as a trace mineral supplement in the feeds for livestock and poultry to prevent copper deficiency diseases. Copper hydroxide in ammonium hydroxide solution is used as a solvent for cellulose for the manufacture of viscose fabric.

## Detection methods
Colorimetric methods are very sensitive for the determination of copper. Reagents such as dithizone, diethylthiocarbamate or salicyladoxime will measure copper at ng/ml concentrations.

Fluorimetric analysis using 1,1,3-tricyano-2-amino–1-propene can be used to measure copper in biological materials.

Polarography using anodic stripping voltammetry on biological materials pre-treated by dry ashing of the sample, dissolving in nitric acid and buffered with acetate buffer to pH 4.3. Under these conditions, copper, cadmium, lead and zinc, can be measured in the same sample at ng/ml concentrations.

Atomic absorption spectroscopy probably offers the best methods for analysis of copper. At the copper resonance line at 324.8 nm, copper containing solutions can be determined using air acetylene flame atomic absorption, flame emission methods, or flameless graphite furnace techniques which are ten times more sensitive than the flame methods.

### Precautions when using the substance
Workers exposed to copper dust when working with copper containing ores must wear appropriate face masks to minimize the inhalation of copper laden dust. Work areas where significant amounts of copper fumes might arise must be well ventilated, if necessary with a mechanical exhaust ventilation system. Metal brazing, welding, or molten metal work will require the provision of a fresh airline hood if the work is carried out in confined spaces.

Spraying operations involving copper fungicides or anti-fouling agents must be carried out using chemical resistant protective clothing and face masks. Ingestion of solutions containing copper salt, which may be toxic, usually results in nausea and vomiting.

### Occupational health
Most workers are virtually immune to poisoning from copper. Although the element is an essential trace metal for normal human metabolism, the amount absorbed from the gastrointestinal tract is controlled to about 5 mg per day. In 1 in 200 000 of the population, an abnormal gene exists which interferes with the normal copper absorption control mechanism and allows excessive amounts of copper to be absorbed and deposited in the liver, brain and kidneys. This disease of copper toxicosis, known after its founder Dr Kinnear Wilson, is progressive and fatal if not treated. The condition can be identified by a reduction in the copper-alpha 2 globulin complex, known as caeruloplasmin, present in the blood serum. (Wilson's disease cases, <20 mg caeruloplasmin/100 ml, compared with normal subjects 20–50 mg caeruloplasmin/100 ml.)

The caeruloplasmin test should be carried out on all workers who are suspected of having this abnormal gene if they are to work in copper contaminated areas.

### First aid

#### Inhalation

Workers who have been exposed to copper containing fumes or dusts must be removed from the contaminated area into fresh air. The effects of exposure are usually seen some 8–10 hours later and usually described as metal fume fever. This condition resembles influenza with general malaise, headache, high temperature and sometimes pain or tightness in the chest. Sweating may follow and the patient will feel extremely tired and sleepy. They should be kept warm and rest in bed. Recovery without further medication occurs in 48 hours.

#### Ingestion

**The ingestion of copper salts, particularly copper sulphate, is a serious matter since most copper salts are toxic and corrosive to the body.** The patient will complain of severe nausea, usually with vomiting, and have intense abdominal colic. The victim should be given milk to drink to buffer the stomach contents, reducing the corrosive nature of the copper salt and delaying the passage of the stomach contents into the duodenum. The victim must be taken to hospital without delay for a gastric lavage to remove any remaining copper salts. Inform the medical staff at the emergency department of the type of copper salt ingested and if possible the amount swallowed, so that further specialist treatment can be given.

#### Skin contact

Copper salts which come into contact with the skin can be easily removed by washing with a stream of cold tap water for at least 10 minutes. If the skin is damaged due to corrosive action of the copper containing chemical, the affected skin area should be covered with a dry dressing and the victim referred to hospital for further specialist treatment.

#### Eye contact

Any concentrated copper solution in contact with the eye must be considered serious. Immediately, irrigate the eye(s) with a stream of cold tap water from a flexible hose to remove as much of the copper salt as possible. Ensure during this irrigation process that the surface of the eye and the eye orbit are thoroughly washed. Place a dry eye pad(s) over the treated eye(s) **and immediately take the casualty to hospital for further specialist treatment.**

# Diborane $B_2H_6$

## Boron Hydride

## Boroethane

**Exposure limits**
OES = 0.1 ppm (0.1 mg/m$^3$)
STEL = none

## Description
**Colourless, highly toxic and flammable gas** with a stickly-sweet nauseating odour. Usually supplied in a mixture with other gases. **It ignites spontaneously in moist air** to give hydrogen and boric acid.

## Properties
MW = 27.7,  SG = 0.45 (liquid),  MP = –165.5°C,  BP = –92.5°C,
VD = 0.96,  VP = 224 mmHg (29.8 × 10$^3$ Pa) at –112°C,
EL = 0.9 – 98%,  FP = –90°C,  IT = 38–52°C.

## Detection methods
Chemical reaction tubes.
Infra-red analyser.

## Precautions when using the gas
**Because of its high toxicity, diborane should not be used without a detailed COSHH assessment and after seeking expert advice on its handling and storage.**

Diborane containers should be refrigerated to prevent decomposition of the gas. Storage, particularly of large quantities of the gas, should be sited in well-ventilated storage areas marked with warning signs. All staff working in storage areas must wear personal protective clothing and full respiratory protection designed for this toxic gas.

The lowest concentration of diborane detectable by smell is 3.3 ppm, which is 33 times greater than the Occupational Exposure Standard. Because of this, monitoring of the atmosphere of the workplace for this gas should be by automatic leak detectors and alarms. Operation of diborane containing apparatus should be by remote control.

**Fires involving diborane are extremely difficult to extinguish.** Small fires can be dealt with using an inert substance such as dry sand or liquid nitrogen if available. Halon extinguishers must never be used as boranes form explosive compounds with halogenated hydrocarbons.

## Occupational health
Inhalation of small quantities of the gas can produce headaches, nausea, muscular fatigue and weakness. Acute poisoning leads to respiratory distress, coughing, pulmonary oedema and haemorrhage, and, if untreated, death. Subacute poisoning

gives rise to signs and symptoms of respiratory damage and intoxication which can persist for some days.

---

**First aid**

Remove victim from the contaminated area into fresh air using full protective clothing and wearing a compressed air breathing set, fitted with a full face mask. Ensure a clear airway; if breathing has stopped or is weak, give artificial respiration using an oxygen resuscitator (not mouth to mouth). If the victim's clothing is contaminated, remove it and then wash the patient's skin thoroughly with cold water for at least 15 minutes. Lie the patient down preferably on a stretcher keeping them at rest and warm. Remove to hospital without delay for expert medical care informing the ambulance staff that the patient has been exposed to diborane.

---

# Dimethyl ether    $(CH_3)_2O$
# Diethyl ether    $(CH_3CH_2)_2O$

Methoxy methane,

Methyl oxide

Ethoxyethane, Ethyl oxide

**Exposure limits**

Dimethyl ether   No OES data available

Diethyl ether   OES = 400 ppm (1200 mg/m³)

STEL = 500 ppm (1500 mg/m³)

## Description
All these substances form a colourless, highly flammable gas at room temperature and possess a sweet ethereal smell. All have anaesthetic properties.

## Properties
**Dimethyl ether** is a colourless gas at room temperature, which has been used as a refrigerant gas, and as an engine and rocket fuel.

MW = 46,   SG = 0.66,   MP = −138°C,   BP = −24.9°C,
VD = 1.62,   FP = 41°C,   IT = 350°C.

**Diethyl ether** is a volatile liquid which is used as an anaesthetic, and in the chemical industry and research as a solvent (dry etherial solution), and as the extracting phase in two phase solvent extraction procedures.

MW = 74.1,   SG = 0.71,   MP = −116.2°C,   BP = 34.6°C,   VD = 2.56,
VP = 442 mmHg (58.8 × 10³ Pa),   FP = −27°C,   EL = 1.8–36%,   IT = 215°C

## Detection methods
Gas-liquid chromatography of an air sample injected into a chromatographic column using a gas injection syringe.
Chemical reaction tubes.
Flammable gas detector.

## Precautions when using the gas
**These ethers are very flammable and have low flash points. Because of these factors great care should be taken to ventilate the workplace to reduce the air concentration of the vapour and avoid accidental ignition.** This is particularly important in workplaces or clinical areas where ether is used in the presence of electrical apparatus.

Storage of diethyl ether solutions can give rise to the formation of ether peroxide which is explosive. If dry ether is not required, peroxide formation can be inhibited by adding 10% water or 5% aqueous ferrous sulphate solution.

## Occupational health

The high value quoted for the OES of diethyl ether shows that the substance is not particularly hazardous to man. The main consideration with regard to working with diethyl ether liquid or vapour is its anaesthetic properties. Inhalation of air concentrations in excess of the OES can give rise to drowsiness or collapse. Diethyl ether liquid is a very good fat solvent and prolonged contact with the skin can give rise to skin cracking and dermatitis. The wearing of gloves and eye protection is essential when handling quantities of liquid ether.

---

**First Aid**

No special precautions are needed when handling casualties with exposures to ether. The subject should be removed from the contaminated area into fresh air, and allowed to recover, if necessary, in the unconscious position.

---

# Ethylene

# $CH_2=CH_2$

## Ethene

### Exposure Limits

OES is not quoted as compound is considered
an asphyxiant by reducing the oxygen content of
respirable air.

## Description
A colourless, flammable gas with a characteristic sweetish smell.

## Properties
Ethylene gas is used in a large number of organic synthetic reactions, the production of polyethylene, and as a refrigerant gas, anaesthetic, and as a fruit ripening agent. Ethylene is insoluble in water, slightly soluble in ethyl alcohol and very soluble in ether.

MW = 28.05,  SG = 0.61,  MP = −169.1°C,  BP = 104°C,  VD = 0.98,
EL = 3.1–32%,  IT = 450°C.

## Detection methods
Chemical gas detector tubes.
Flammable gas detection equipment.

## Precautions when using the gas
**The greatest dangers from ethylene are fire and explosion**. The gas concentration in the working environment should not exceed 0.6% by volume (20% of the Lower Explosion Limit LEL). This must be achieved by good ventilation using electrical fans and equipment designed to be explosion proof. Tests of the working atmosphere must be made using appropriate flammable gas detection equipment designed to be used in flammable gas atmospheres and to indicate the % LEL of the gas/air mixture present.

**In industrial processes where liquid ethylene is used, careful planning is required to protect staff and equipment in the event of fire.** Fires involving ethylene are extremely difficult to extinguish, particularly in pipeline installations. Where possible static water spray facilities should be available which will operate automatically in case of fire, and the supply of ethylene gas into the pipeline should be shut off allowing the fire to burn itself out.

## Occupational health
Ethylene gas acts mainly as an asphyxiant by reducing the oxygen content of respirable air. In high concentrations ethylene acts as an anaesthetic and may render the subject drowsy or unconscious. There is no medical evidence that prolonged exposure to low concentrations of ethylene produces chronic or adverse effects.

**First aid**

Inhalation

Remove the subject from the contaminated area into fresh air, the rescuers wearing compressed air breathing apparatus, not canister respirators because of low oxygen availability. If the subject is breathing normally allow to recover at rest and keep warm to reduce shock. If the victim has difficulty in breathing, give 100% oxygen or oxygen-enriched air from a resuscitator until normal breathing is restored. **The casualty should be seen by a medically qualified person as soon as possible.**

Skin contact

Liquid ethylene can easily produce a cryogenic burn or frostbite. Immerse the affected area in warm tap water (42°C) if available. If not, cover the affected area with woollen blankets or coats to allow the circulation in the affected part to be restored gradually. **The casualty should be taken to hospital without delay for expert medical treatment.**

Eye contact
**LIQUID ETHYLENE SPILLAGE INTO THE EYE IS A MEDICAL EMERGENCY.**

Immediately irrigate the eye with a steady stream of cold tap water to remove as much ethylene as possible and to slowly warm the eye and the surrounding tissues. Continue treatment for at least 15 minutes. Call an ambulance and arrange for the casualty to be taken to hospital for further specialist medical treatment.

# Ethylene oxide $C_2H_4O$

## 1,2-Epoxyethane

**Exposure limits**
OES = 5 ppm (10 mg/m$^3$)
Any long-term exposure should be reduced to a minimum in view of animal exposure experiments indicating possible carcinogenicity.

## Description
Ethylene oxide is a colourless, flammable gas at room temperature possessing an ether-like smell.

## Properties
The gas is soluble in water, ethyl alcohol, and diethyl ether. It is miscible with most aqueous and organic solvents. It is used in the production of ethylene glycol, polyethylene, and as a fumigant and sterilizing agent for heat sensitive medical equipment.

MW = 44.05,　SG = 0.87,　MP = –112°C,　BP = 10.4°C,　VD = 1.5,
VP = 1095 mmHg (142.3 × 10$^3$ Pa),　FP = –6°C,　EL = 3–100%,　IT = 429°C.

## Detection methods
Infra-red gas analysis.
Chemical reaction tubes.

## Precautions when using the gas
Because of the relatively low Occupational Exposure Standard and the flammability of the gas, great care must be taken to avoid the build-up of vapour in the workplace from leakages. Ignition of the vapour can occur easily even with static electricity. It is essential that all workplaces using ethylene oxide must have high flow exhaust ventilation. Spillages of the liquid or releases of significant volumes of the gas require proper emergency procedures and availability of chemical resistant clothing and respiratory protection equipment.

## Occupational health
Reports of accidental high level exposures to ethylene oxide show that the gas causes acute irritation to the eyes, nose and throat resulting in coughing and vomiting. Delayed effects may produce pulmonary oedema, bronchitis, and emphysema. Rapid evaporation of the liquid can produce frostbite. Skin irritation is caused by prolonged contact with ethylene oxide contaminated gloves or clothing. This can be avoided by ensuring all contaminated clothing is removed and the skin carefully washed. Ethylene oxide can form ethylene chlorhydrin in the presence of moisture and chloride ions. The chlorhydrin is a human systemic poison and exposure to it can produce fatal results.

**First aid**

Workers exposed to high concentrations of the vapour may show signs of drowsiness. They should be removed form the contaminated area into fresh air. Make sure that the subject's airway is clear. If the patient's breathing is poor or has stopped, give artificial respiration preferably with an oxygen-supplemented resuscitator.

# Ethyl bromide $C_2H_5Br$

## Exposure limits
OES = 200 ppm (890 mg/m³)
STEL = 250 ppm (1110 mg/m³)

## Bromoethane

## Description
A volatile liquid which produces an ethereal smelling vapour which decomposes on exposure to oxygen and light.

## Properties
The liquid is sparingly soluble in water.

MW = 109,   SG = 1.45,   MP = –119°C,   BP = 38°C,   VD = 3.76,
VP = 400 mmHg (532 × 10³ Pa) at 21°C,   FP = –16°C,
EL = 6.7–11.3%,   IT = 511°C.

## Detection methods
Gas chromatographic analysis.
Infra-red spectroscopy.

## Precautions when using the gas
The vapour from ethyl bromide is less toxic than methyl bromide but has greater general toxicity than ethyl chloride. Ethyl bromide should not be used or handled in unventilated areas or workplaces. Its vapour can irritate the respiratory system and the eyes. The liquid should not be ingested as it can affect the kidneys and nervous system.

## Occupational health
Exposure to ethyl bromide produces noticeable irritation to the respiratory system. Concentrations of 10% can lead to narcosis: the vapour has been used in the past as a general anaesthetic but this practice is no longer accepted. Prolonged acute industrial exposure to the vapour has been shown to lead to damage to the extra pyramidal nervous system, heart and to liver degenerative lesions. Chronic exposures give rise to dizziness and weakness in the limbs, giving rise to disturbances in walking, hyperreflexia and speech defects.

**First Aid**

Inhalation

Concentrations of the vapour >3% can give rise to unconsciousness and loss of coordination. Persons exposed to ethyl bromide should be removed from the contaminated area to fresh air, and allowed to recover under supervision of a trained first aider. Inhalations of high concentrations of the vapour may require oxygen-enhanced artificial respiration for the victim. The casualty should then be transferred to hospital for further medical treatment and observation as complications may occur.

Skin contact

Any clothing contaminated with the liquid should be removed and the affected area washed thoroughly with soap and rinsed with large volumes of cold tap water. Spillages onto the skin can result in skin absorption of the liquid and the casualty should be seen by a medical practitioner or taken to hospital following the first aid treatment.

Eye contact

**Splashes of liquid into the eyes should be treated as a serious matter.** Immediately treat the casualty by irrigating the eye with a slow stream of cold water to prevent further tissue damage and absorption of the chemical. The irrigation should continue for at least 10 minutes.

Cover the eye with a dry eye dressing pad (do not place any medication into the eye(s)) **and take the casualty to hospital without delay**. Inform the medical staff at the hospital of the nature of the exposure and the substance involved.

# Ethyl chloride $C_2H_5Cl$

Chloroethane

**Exposure limits**
OES = 1000 ppm (2600 mg/m³)
STEL = 1250 ppm (3250 mg/m³)

## Description
A colourless gas with a characteristic ethereal smell. **This compound is much less toxic than methyl chloride but has fire and explosion risks.**

## Properties
Ethyl chloride is usually supplied in cylinders as a liquefied gas. The gas is sparingly soluble in water.

MW = 64.5, SG = 0.92, MP = –139°C, BP = 12.3°C, VD = 2.22,
VP = 1000 mmHg (133 × 10³ Pa) at 20°C, FP = –50°C, EL = 3.8–15.4%,
IT = 510°C.

## Detection methods
Gas chromatographic analysis.
Infra-red spectroscopy.

## Precautions when using the gas
Although the gas is not particularly toxic **it should not be used in unventilated areas or basements as leakages from gas cylinders containing ethyl chloride can reduce the oxygen content of breathable air.** Skin contact with the liquid from cylinders will produce cold burns or frostbite.

Although ethyl chloride liquid is used as a freezing anaesthetic for minor operations on the skin, it should not be used as an inhalation anaesthetic. There have been reports that ethyl chloride if inhaled and absorbed will markedly depress the respiratory and circulatory system of the body.

## Occupational health
The vapour at room temperature is relatively non-toxic, acting as an anaesthetic in high concentration in air. The vapour can irritate the mucous secreting epithelial cells of the lungs especially if there is any possibility of thermal decomposition of ethyl chloride to produce an admixture of hydrochloric acid gas.

**First aid**

Inhalation

Concentrations of 2–4% of the gas can cause loss of consciousness and coordination. Persons exposed to ethyl chloride gas should be immediately removed to fresh air and allowed to recover under supervision of first aid trained staff. Large or massive exposures may require oxygen-enhanced artificial respiration.

Skin contact

The greatest danger to the victim of gas or liquid contact is the rapid freezing action leading to a cold burn or frostbite. Where skin is concerned any clothing saturated with the gas or liquid should be removed and the affected area drenched with tap water (if available immerse in tepid, not hot, water until the blood flow at the surface of the skin appears normal). If the burn is severe, cover with a dry loose dressing and refer the casualty to a hospital or to the nearest medically qualified doctor for further treatment.

Eye contact

**Splashes of liquefied gas into the eye is a serious matter.** Immediately treat the casualty by irrigating the eye with a stream of cold tap water to prevent further tissue damage. The irrigation should be continued for at least 10 minutes. Cover the eye with a dry eye dressing, do not place any medication in the eye **and take the patient to hospital without delay**. Inform the medical staff at the hospital of the nature of the exposure and the substance involved.

# Fluorine  $F_2$

**Exposure limits**
No long-term OES quoted
STEL = 1 ppm (1.5 mg/m³)

## Description
A yellow-coloured, violently reactive, highly toxic and irritant gas.

## Properties
Fluorine is the most electronegative element, and exists both as a liquid and gas. In either form it is highly dangerous and must be handled with extreme care and with the appropriate safety equipment.

Fluorine decomposes water with the production of hydrofluoric acid and ozone. It explodes on contact with hydrogen, and reacts violently with ignition in the presence of iodine, sulphur, alkaline metals and many organic chemicals.

MW = 38,  SG = 1.11 (liquid),  MP = −219.6°C,  BP = −188.1°C,
VD = 1.69,  VP = >1 bar.

## Detection methods
Leaks from pipework can be detected with ammonia vapour producing dense white ammonium fluoride. Filter paper moistened with a solution of potassium iodide will detect fluorine gas concentrations >25 ppm in air.

## Precautions when using the gas
When using any quantity of fluorine, it is essential that before any work is done, that a detailed risk assessment is made of the process to be used and the appropriate safety measures provided.

Processes involving fluorine should be set up in a separate well-ventilated area which can be isolated in the event of a serious leakage of liquid or gas. In industrial plant where fluorine gas may be used in pipework under pressure the dangers are much greater. Provision must be made for the control of leakage or the presence of fire in the fluorine plant area. The installation of water mist drenches may be necessary to deal with these events.

Storage of fluorine cylinders must be in well-ventilated lockable steel cages in as isolated area as possible.

All equipment used near to fluorine containing equipment must be gas tight to reduce corrosion. Any plant used for fluorine must be carefully cleaned, degreased and dried, and then fluorine introduced with increasing concentration so that any impurities present in the apparatus can be burnt out under controlled conditions without ignition of the whole plant.

Valves fitted in pipelines should be remote controlled either by use of long mechanical extension handles or electrically. Pressure control of the pipeline must be by a two-stage valve control system to minimize accidents caused by failures.

All staff must be instructed in the hazardous nature of the substance and fully trained in the safety measures and personal protective equipment needed* when dealing with normal work, and in the event of an accidental spillage of the liquid or discharge of gas. In factories, it is essential to provide fast acting safety showers, drenches and eye wash stations.

## Occupational health

Occupational health data suggest that no long-term hazards will result if workers are not subjected to levels of fluorine gas >0.1 ppm in air. At this concentration the gas produces mild irritation to the respiratory tract. Inhalation of higher concentrations at the workplace of the order of the STEL = 1 ppm can produce irritation and damage to the respiratory tract, pulmonary oedema, haemorrhage and kidney damage.

---

*The level of personal protection needed will depend on the scale of operations carried out. For small laboratory work, this would require neoprene apron and long gloves and boots worn in conjunction with a chemical resistant overall, and face and eye protection. For large industrial operations, a fully enclosed chemical resistant suit with airline hood would be suitable. Staff working with this type of dangerous material must not work alone.

### First aid

**Exposure to fluorine, particularly the liquid form, must be considered a medical emergency. The presence and/or advice from a medically qualified person is required as soon as possible.**

### Inhalation

The subject should be removed to fresh air by rescuers wearing breathing apparatus of an approved type, and given 100% oxygen, preferably by means of an oxygen resuscitator. Even with minor exposures, 100% oxygen should be administered at intervals for up to 4 hours following exposure. In severe exposures, the victim should receive 100% oxygen therapy immediately, **and then removed by ambulance to hospital without delay**. A member of the staff should accompany the casualty to hospital to give details to the medical staff of the type and duration of the exposure, to assist in the patient's further treatment.

### Eye contact with vapour or liquid

Without delay, irrigate the eyes and face with large volumes of cold tap water for at least 15 minutes. Get an assistant to call for an ambulance. Do not leave the patient. Repeat the irrigation treatment if necessary. The victim should be removed to hospital and seen by an ophthalmic specialist.

### Skin contact

Immediately drench the affected area or whole body with a shower of cold water. Remove the victim's clothing while under the water shower. Continue cold water drenching for at least 10 minutes to remove as much fluorine from the skin. **Get an assistant to seek medical assistance as soon as possible**.

Treat the affected area with iced 70% ethyl alcohol (ethanol) or an iced solution of magnesium sulphate for 30 minutes. If medical aid is not available, continue the ethanol or magnesium sulphate treatment for 2 hours, and then apply a paste of magnesium oxide in glycerine. Alternatively, use a liberal application of calcium gluconate gel to the affected area.

Note: Some medical practitioners prefer subcutaneous injections of 10% calcium gluconate solution, but the use of calcium gluconate gel has proved to be effective in treating liquid splashes and burns.

Administer pure oxygen, or give artificial respiration with supplementary oxygen if breathing is weak. Keep the patient warm and at rest.

Modern acute medical treatments consist of exchange transfusions, mannitol diuresis and urinary alkalization and possible peritoneal or haemodialysis if renal failure occurs.

# Formaldehyde ## HCHO

## Methanol

**Exposure limits**
OES = 2 ppm (2.5 mg/m$^3$)
STEL = 2 ppm (2.5 mg/m$^3$)

### Description
A colourless relatively toxic gas with a characteristic pungent odour at room temperature.

### Properties
The gas is very soluble in water, alcohol and ether. It can form concentrated solutions in these solvents that can be used for preservation and as a bactericide and fungicide.

Formaldehyde and its solutions are flammable and the vapour when mixed with air can be explosive. It can easily polymerize into a tetramer and a polymer, paraldehyde.

MW = 90.08,  SG = 1.0,  MP = –92°C,  BP = –21°C,  FP = 50°C,
EL = 7–73%,  IT = 430°C.

### Detection methods
Gas detection fuel cell specific for formaldehyde gas at the OES range.
Chemical titration methods.
Chemical gas detector tubes.

### Precautions when using the gas
Because of the low OES set for this compound, great care must be exercised in maintaining a high level of extract ventilation in areas where formaldehyde solutions, the solid polymeric forms or vapour are used. Protective clothing is required whenever formaline solutions or formaldehyde are handled. This is particularly important in pathology and anatomy departments where formaldehyde is used to preserve and treat infected tissue specimens. Since the STEL is 2 ppm formaldehyde in air, it is difficult to meet this standard at all times in the workplace. Both high and low level ventilation is required in laboratories and workplaces where this substance is being used on a regular basis. **All work with the vapour must be carried out in an area fitted with high flow air extraction or in an exhaust protective cabinet with a flow rate not less than 1 metre/sec. across the working face of the cabinet. In industrial workplaces, where high concentrations of vapour or solutions are handled, chemical resistant clothing and respiratory protection must be used and the work thoroughly supervised.**

## Occupational health

Formaldehyde is toxic by ingestion and inhalation and contact with the skin can lead to epithelial damage due to protein denaturation, and in low concentration can give rise to allergic reactions. There is strong evidence from animal work that formaldehyde is a potential carcinogen hence the low Occupational Exposure Standard set by the HSE. Formaldehyde vapour is detectable by smell at concentrations below 1 ppm, although olfactory fatigue can be developed during prolonged low concentration exposure. Concentrations of gas as low as 3 ppm can cause irritation of mucous membranes, watering of the eyes, and a cough. The presence of perceivable formaldehyde in the air is sufficient evidence that air extraction and control measures are not adequate. **In factories where both formaldehyde and hydrogen chloride gas are used, great care must be exercised to ensure that these substances are not allowed to react. If this occurs a highly dangerous carcinogen, bis-chloromethyl ether, can be produced.**

---

**First aid**

Inhalation
The subject must be removed from the contaminated area into fresh air as soon as possible, the rescuers wearing breathing apparatus. If breathing is normal and the subject appears well oxygenated, keep warm and at rest, and allow to recover normally. If there is any difficulty with breathing, give oxygen-enriched air from an oxygen resuscitator until the patient's breathing has returned to normal.

Skin contact
Any contact with the skin should be treated by washing the affected area with cold water and soap until the smell of formaldehyde is removed. If the affected area is very large or the solution in contact with the skin is concentrated, the patient should be taken to hospital for further treatment.

Ingestion
If it is known that the subject has swallowed formaldehyde, they should be given 0.5 litres of milk or treated with 0.5 litres of water containing ammonium acetate, to inactivate the formaldehyde. Vomiting should be induced to prevent absorption of the formaldehyde via the stomach. The patient should be removed to hospital by ambulance, the ambulance staff being told of the type of poisoning suspected.

Eye contact
Immediately irrigate the eye and surrounding tissues with copious quantities of cold tap water, making sure that the eye and the eye socket are fully decontaminated. Continue this process for at least 15 minutes. Place a dry eye dressing over the eye, **and take the patient to hospital without delay to obtain ophthalmic medical treatment**.

---

# Germane

# GeH$_4$

## Germanium tetrahydride

**Exposure limits**
OES = 0.2 ppm
STEL = 0.6 ppm

### Description
A toxic, flammable gas with a characteristic pungent odour.

### Properties
It is used extensively in the electronics industry in the preparation of pure germanium crystals in the manufacture of superconductors. Germanium hydride and chloride are considered the most toxic germanium compounds.

MW = 76.6,  SG = 1.52 (liquid),  D = 3.43 (gas),
MP = −165°C,  BP = −88.5°C.

### Precautions when using the gas
**Germane is a very toxic gas and the minimum lethal dose for man is unknown**. The precautions taken when using germane are very similar to those for arsine (arsenic hydride). **Any work with germane must be carried out in a specially ventilated laboratory or workshop in an exhaust protective cabinet fitted with a filter to remove germane from the exhaust air. No work with germane should be sanctioned until a full and rigorous safety risk assessment has been prepared and the safe working arrangements made and tested**.

Germane is usually supplied in small cylinders which must be stored in a locked, ventilated store. **In the event of a leakage of gas, trained staff, using positive pressure breathing apparatus, should attempt to shut the leaking cylinder valve or move the cylinder into a filter cabinet or to the open air, downwind from staff and buildings. The incident should be immediately reported to the fire brigade.**

### Occupational health
Fortunately poisoning from germane is rare but the gas should be considered as a potential haemolytic and nephrotoxic agent. Precautions similar to those given for arsine would be appropriate.

Because germanium is near to lead in the Periodic Table of elements, it forms stable and inert chelates with EDTA (ethylene diamine tetraacetic acid) or calcium disodium versenate. These germanium chelates can be excreted in the urine and could form an effective treatment of germanium intoxication in man.

**First aid**

GERMANE POISONING OR INTOXICATION IS A MEDICAL EMERGENCY.

The victim should be removed from the contaminated area by a team of rescuers wearing protective clothing and compressed air breathing apparatus. If the patients breathing is impaired, 100% oxygen should be administered using a resuscitator. **On no account should mouth-to-mouth resuscitation be used. The patient must be removed to hospital by ambulance without delay, preferably with medically qualified or paramedical staff in attendance, and a member of staff from the workplace to explain the nature of the poisoning to the hospital emergency staff. The hospital receiving this patient should if possible be informed by telephone of the nature of the accident so that appropriate medication can be organized.**

# Halothane $CHBrCl-CF_3$

**Exposure limits**
OES = 10 ppm (80 mg/m$^3$)

## Description
A colourless, non-explosive, halogenated hydrocarbon gas. The gas burns and may release fluorine, bromine or chlorine or their oxidation products.

## Properties
Halothane is used as a clinical anaesthetic, concentrations of the gas up to 5% are used to induce anaesthesia; the gas concentration needed to maintain anaesthesia is much lower, 1–1.5% usually being satisfactory.

## Detection methods
Gas – liquid chromatography.
Ultraviolet and infra-red spectroscopy.

## Precautions when using the gas
Adequate ventilation is essential to avoid the build-up of narcotic concentrations. At low concentrations of the gas, accumulation occurs in the body because of the high solubility of halothane in the body's fat stores. Metabolism of halothane takes place slowly in the body following exposure. Driving or handling of machinery should be avoided for 24 to 48 hours following exposure that has produced anaesthesia.

## Occupational health
**Anaesthesia is associated with the lowering of blood pressure related to the concentration of the gas inhaled.** The low blood pressure is related to the depression of the sympathetic nervous system, myocardial depression and the loss of tone in the smooth muscles, including those in the peripheral vascular system.

**First Aid**

Persons who have received an overdose of halothane gas should be carried immediately to fresh air, place on the ground or a stretcher and carefully treated by artificial ventilation preferably with oxygen supplementation until the patient recovers. They should be kept warm and at rest. **They should not be left unattended in case they lapse into unconsciousness and need further resuscitation. Patients should be seen by a medically qualified person as soon as possible to decided what further treatment is required.**

Note: Drugs that raise blood pressure by vasoconstriction are contraindicated as they are liable to cause cardiac fibrillation or arrest.

# Hydrocarbons (aliphatic)

Olefines

Ethylene (Ethene) see under
Ethylene

Propylene (Propene)

Butylene (Butene)

**Exposure limits**
OES = not quoted as the gas is considered as an
asphyxiant by reducing the oxygen content of
respirable air

### Description
Both propylene and butylene are colourless, flammable gases used in the polymer industry in large quantities.

### Properties
**Both gases are highly flammable and explosive if mixed with air or oxygen.**

Propylene
MW = 42.1, D = 0.58, MP = −185°C, BP = −47.8°C, VD = 1.5,
VP = 7600 mmHg (1011 × $10^3$ Pa), FP = 108°C, EL = 2–11.1%,
IT = 497°C.

Butylene
MW = 56.1, D = 0.66, MP = −185.3°C, BP = −6.3°C, VD = 1.9,
VP = 3480 mmHg (462.8 × $10^3$ Pa), FP = 108°C, EL = 1.6–9.3%,
IT = 384°C.

### Detection methods
Fuel cell type flammable gas detector.

### Precautions when using the gas
As previously stated, **these gases are extremely flammable and explosive when mixed with air or oxygen. It is recommended that the working atmosphere should be continuously monitored for oxygen concentration to ensure this level remains above 18%. Monitoring for the presence of these flammable gases should be made so that the concentration of flammable gas in the air does not exceed 0.32% of the lower explosion limit.**

### Occupational health
Gaseous aliphatic hydrocarbons have a low toxicity. In high concentrations they act as asphyxiants due to their effective reduction of available oxygen in contaminated air. In very high concentrations of the order of 60%, ethylene and propylene have

been used as surgical anaesthetics but in this role they must be administered with medical oxygen.

---

**First aid**

## Inhalation

**Remove the casualty from the contaminated area into fresh air, the rescuers wearing compressed air breathing apparatus.** If the subject is breathing normally allow to recover but be ready to give oxygen if breathing becomes distressed. If the casualty is unconscious, give artificial respiration preferably using an oxygen-supplemented resuscitator. A medically qualified person should give advice on any further treatment required as soon as possible. If in doubt, call an ambulance and remove the casualty to hospital for further assessment.

## Skin contact

**Liquefied ethylene, propylene or butylene can easily produce a cyogenic burn or frostbite.** If the liquid has contaminated the skin, remove any contaminated clothing, and if possible, immerse the affected area in warm tap water (42°C), or cover the area with woollen clothing or blankets to allow the blood circulation to the tissues to be gradually restored to normal. **The casualty should be taken to hospital without delay.**

## Eye contact

Irrigate the eyes with a steady stream of cold tap water for at least 15 minutes making sure that the eye and surrounding tissues are fully washed. This will allow the tissues to warm up slowly and reduce tissue damage. Cover the eye(s) with a dry eye pad dressing **and refer the casualty to the nearest hospital Accident and Emergency department for further specialist medical treatment.**

---

# Hydrogen

# H$_2$

## Exposure limits
No safe limits of exposure are quoted.

## Description
A highly flammable, colourless, odourless gas, lighter than air. Mixtures of hydrogen with air are explosive.

## Properties
The gas is slightly soluble in water.

MW = 2,  SG = 0.09,  MP = −259.1°C,  BP = −252.5°C,  VD = 0.07, EL = 4.1–74.2%,  IT = 400°C.

## Detection methods
Flammable gas detectors.
Special chemical detector tubes.

## Precautions when using the gas
Hydrogen gas in any concentrations with air or oxygen is extremely hazardous. Liquefied hydrogen presents even greater risks from fire and explosion because of its low evaporation point, small quantities of heat can generate large volumes of potentially explosive gas. Any oxygen or air entering liquid hydrogen storage tanks can be readily solidified and the build-up of solid air or oxygen particles in the liquid hydrogen can trigger a violent explosion. Tanks being used to store liquid hydrogen must be purged with dry nitrogen gas before filling

Hydrogen gas/air mixtures are easily ignited by electric sparks or discharges. Leaking cylinders present a danger of fire and/or explosion in the workplace. Before using any hydrogen gas cylinder or flexible gas lines, test the tightness of all joints by using soap solution. If the cylinder is found to have a slow leakage of gas, remove all sources of ignition from the area, and then remove the cylinder from the building into the open air, allowing it to vent slowly to atmosphere, preferably downwind from the building, placing warning signs and sealing off the affected area.

In the event of a major leak of gas and/or fire, call the fire brigade, evacuate the area and leave staff at a safe distance to prevent others entering the danger zone. (Remember fuel gas cylinders can explode in fires.)

## Occupational health

Hydrogen is considered non-toxic but at high concentration would act as an asphyxiant as it would deplete available oxygen in the air.

---

### First aid

Inhalation

Remove the victim from the contaminated area. If the breathing is seen to be weak or stopped, give oxygen-supplemented artificial resuscitation until the patient appears to breathe normally. Keep the patient warm and at rest.

Skin and eye contact

**Liquid hydrogen presents a serious medical problem if it comes into contact with the skin or eyes.** The first aid treatment is that for a serious cold burn, using copious drenching of the skin or eyes with cold tap water for at least 10 minutes. Expert medical aid is required as soon as possible to decide on further treatment.

Splashes into the eye of the liquefied gas will probably cause permanent damage to the eye because of its extremely low temperature. Cold water irrigation is for at least 15 minutes is essential to minimize further damage to tissues. A dry eye dressing should be used to protect the eye **and the casualty must be taken to hospital immediately following the first aid treatment and seen by an eye specialist without delay.**

---

# Hydrogen bromide HBr

**Exposure limits**
OES = 3 ppm (10 mg/m³)
These values relate to a Short-Term Exposure
Limit (10 minute reference period) only.

## Hydrobromic acid

### Description
Hydrogen bromide exists in both liquid and vapour forms. The concentrated acid is a faintly yellow corrosive and pungent smelling liquid, which darkens on exposure to air and light producing bromine gas. **Both the liquid and the vapour should be treated with caution.**

### Properties
Hydrogen bromide is used for the synthesis of inorganic and organic bromides. The gas is soluble in water and ethyl alcohol.

MW = 80.9,   SG = 3.5,   MP = −88.5°C,   BP = −67°C.

### Detection methods
The presence of hydrogen bromide gas is indicated by the production of dense white fumes of ammonium bromide with 0.88 SG ammonia solution.
Chemical reaction tubes.

### Precautions when using the gas
Hydrogen bromide gas is supplied in cylinders for use in closed circuit systems. The workplace must be well ventilated to prevent any build-up of gas. Storage of gas cylinders should be in well-ventilated areas. Workers involved in using the gas must wear acid resistant clothing, with hand, foot, and eye protection. Respiratory protection such as self-contained compressed air equipment, breathing air line with full face mask or a full face HBr canister respirator must be worn. The choice of breathing protection must be based on a risk assessment of the workplace and the amount of hydrogen bromide used. Provision of emergency showers and/or eye wash stations must be provided where hydrogen bromide is used on a regular basis.

### Occupational health
The most serious health effects caused by contact with hydrogen bromide is its corrosive effects on the skin and mucous membranes. The acid will produce burns on the skin, and the resulting vapours can damage the respiratory tract causing bronchitis and pulmonary oedema.

### First aid

#### Inhalation

Hydrogen bromide vapours or liquid spillage produce choking and coughing in exposed subjects. Remove the victim from the affected area (rescuers wearing respiratory protection) into fresh air. **If breathing is difficult or stops, give artificial respiration preferably using a resuscitator.** Keep the subject warm and at rest. Severe exposure can cause shock, rapid pulse, sweating and collapse. **Medically qualified assistance is required as soon as possible.**

#### Skin contact

**Wash the contamination away from the skin with large volumes of cold tap water,** preferably with a hosepipe or shower unit. **Remove all contaminated clothing from the victim. Cover the affected part with a dry dressing and seek medical advice as to further treatment.**

#### Eye contact

**Irrigate the eyes with large volumes of cold tap water making sure that the recesses of the eyes are thoroughly washed clean of hydrogen bromide.** Do not use jets of water into the eye as this action can drive the acid into the eye socket. Hold the patient's head back and allow a steady stream of water from a flexible hosepipe to irrigate the eye(s). **Irrigation should be continued for at least 15 minutes. The victim should be seen as soon as possible by a medically qualified person for assessment of any further treatment.**

# Hydrogen chloride HCl

## Hydrochloric acid

**Exposure limits**
OES = 5 ppm (7 mg/m³)
These values relate to a Short-Term Exposure
Limit (10 minute reference period) only.

### Description
Hydrogen chloride exists in both liquid and vapour forms. The concentrated acid is a slightly yellow corrosive and pungent liquid which fumes in moist air. **Both the liquid and the vapour should be treated with caution.**

### Properties
The vapour has a characteristic irritating acidic smell even at low concentrations. Hydrochloric acid is soluble in water, ethyl alcohol, and diethyl ether but insoluble in hydrocarbons. Hydrochloric acid solution and vapour are corrosive to most metals.

MW = 37,   SG = 1.19 (38% solution known as 'concentrated'),
MP = −114.8°C (pure HCl),   BP = −84.9°C,   VP = >1 atmos. (>$10^5$ Pa).

### Detection methods
The presence of hydrogen chloride gas is indicated by the production of dense white fumes of ammonium chloride with 0.88 SG ammonia solution.
Chemical reaction tubes.

### Precautions when using the gas
Processes using hydrogen chloride gas should be carried out in closed circuit systems, and the plant situated in well-ventilated areas. Storage of liquid or gas cylinders should be in well-ventilated and covered spaces away from any metal or metal hydrides, oxidizing or flammable materials. Workers handling or using large quantities of acid solutions must wear acid resistant clothing with hand, foot and eye protection. Respiratory protection either as self-contained air breathing equipment or HCl type canister respirators will be required, the choice of equipment dependent on the concentration of the gas present in the workplace. Provision of emergency showers and/or eye wash stations must be provided where hydrogen chloride is used on a regular basis.

### Occupational health
The most serious health effects caused by contact with hydrogen chloride is its corrosive effects on the skin and mucous membranes. The acid will produce burns to the skin, and the resultant acid vapours can damage the respiratory tract causing bronchitis, pulmonary oedema and, in severe cases, death. Prolonged

exposure in the workplace to vapour will produce digestive diseases and a form of dental necrosis in which the teeth become yellow, soft and disintegrate due to the acid attack.

---

**First aid**

Inhalation

Hydrogen chloride vapours produce choking and coughing in exposed subjects. Remove the victim from the affected area into fresh air, the rescuer(s) wearing suitable breathing apparatus. **If breathing is difficult or stops, give artificial respiration preferably using a resuscitator.** Keep the subject warm and at rest. Severe exposure may produce shock, rapid pulse, sweating and collapse. **Medically qualified assistance is required as soon as possible.**

Skin contact

**Wash the soluble hydrogen chloride away from the surface of the skin with a large volume of cold water preferably from a flexible hosepipe or shower unit as quickly as possible. Speed is of the essence to reduce the risk of sustaining severe acid burns.** Remove all contaminated clothing from the victim. Usually, extensive spillage of hydrochloric acid onto clothing causes it to disintegrate. **Continue the water drenching treatment for at least 10 minutes.** Gently dry the surface of the skin with absorbent paper and then take the patient to the nearest Accident and Emergency department for a medical assessment of the injury.

Eye contact

**Irrigate the eyes with a steady stream of cold tap water for at least 10 minutes making sure that the recesses of the eyes have been thoroughly washed out. Do not use jets of water during irrigation as this procedure can force acid into the eye socket and cause further damage. Repeat the irrigation procedure if you think there may be more acid contamination present. Medical advice should then be sought as soon as possible.**

---

# Hydrogen cyanide     HCN
# Cyanogen           $(CN)_2$
# Cyanogen chloride     CNCl

### Exposure limits
Hydrogen cyanide
No OES
STEL = 10 ppm (10 mg/m³)
Cyanogen
OES = 10 ppm (20 mg/m³)
No STEL
Cyanogen chloride
No OES
STEL = 0.3 ppm (0.6 mg/m³)

Prussic acid

Ethane dinitrile, Oxalonitrile

Chlorine cyanide

## Description
**All cyanide containing gases are flammable and very toxic.**

Hydrogen cyanide exists as a colourless liquid or gas with a smell of bitter almonds.

Cyanogen is a colourless gas with an almond odour.

Cyanogen chloride is similar to cyanogen but more toxic, and the vapour is irritating to the eyes and the lungs.

Hydrogen cyanide and cyanogen have similar chemical and biological properties.

## Properties

|  | Hydrogen cyanide | Cyanogen | Cyanogen chloride |
|---|---|---|---|
| MW | 27.03 | 52.04 | 61.5 |
| SG | 0.69 | 0.87 | 1.18 |
| MP | −14.0°C | −27.9°C | −6.9°C |
| BP | 26.0°C | −20.7°C | 12.6°C |
| VD | 0.94 | 1.8 | 2.0 |
| VP | 760 mmHg (25.8°C) | — | 1000 mmHg (20°C) |
|  | (101.3 × 10³ Pa) | — | (133 × 10³ Pa) |
| FP | −17.8°C | — | — |
| EL | 6–41% | 6.6–32% | — |
| IT | 538°C | — | — |
| water solubility | very soluble | very soluble | slightly soluble |
| alcohol solubility | very soluble | very soluble | soluble |
| ethyl ether | very soluble | very soluble | — |

## Detection methods

Hydrogen cyanide and cyanogen – methyl orange/mercuric chloride test papers, or chemical reaction tubes.
Cyanogen chloride – specific chemical reaction tubes.

## Precautions when using the gas

Because of the extreme toxic nature of most cyanides, indicated by the low Occupational Exposure Standards, work involving the use of solid or liquid or gaseous forms of the substance must be strictly controlled, and a system of work devised to assess and eliminate hazards. Safety preplanning is essential especially where large quantities of soluble or gaseous cyanide products are used, e.g. in the electroplating industry. In laboratories where large cylinders of cyanide containing gas are used, procedures for dealing with leaking valves must be devised before any gases are used.

Large leaking cylinders should be removed either by a properly trained and equipped safety team or the Fire Brigade Chemical Incident Unit to the open air and the area sealed off from other persons.

Small cylinders in laboratories can be placed into a tank of cold tap water saturated with calcium hypochlorite (bleaching powder) in a well-ventilated place.

The area where cyanides are to be handled must be defined and very efficiently ventilated either natural or by mechanical means. Where practicable, the process should be completely enclosed, and mechanically ventilated tests of the extracted air should be made to ensure the concentration of cyanide is below 2 ppm otherwise gas scrubbing of the extracted air to remove cyanide will be necessary.

Workers using cyanide must wear approved protective clothing and respiratory protection to prevent cyanide being inhaled or absorbed into the body. Cyanide salts that could release cyanide containing gases must be covered in secure drums or kept in ventilated areas. All cyanide work and storage areas should be separated from other workplaces and clearly marked. Workers must be told of the dangers of allowing cyanide material to come into contact with acids, and should be able to recognize the almond smell of cyanides.

### Requirements for respiratory protection when using hydrogen cyanide gas

| HCN concentration (ppm) | Requirement for respiratory protection |
| --- | --- |
| <90 ppm | (a) Type C supplied air respirator fitted with continuous flow or demand valve half or full face mask. (b) Full face gas mask fitted with canister specific for cyanide. (Maximum service life is 1 hour.) |
| <200 ppm | Full face gas mask fitted with canister specific for hydrogen cyanide. (Maximum service life is 1 hour.) |
| >200 ppm | Self-contained positive pressure breathing apparatus or positive pressure breathing air line with auxiliary self-contained air supply fitted with full face mask worn under gas tight suit providing whole body protection. |
| Emergency rescue work | Positive pressure self-contained breathing apparatus or positive pressure breathing air line with auxiliary self-contained air supply fitted with full facepiece and worn under a gas tight suit providing whole body protection. |

| Evacuation or escape | Self-contained breathing apparatus in demand or positive pressure mode or gas mask full face or mouthpiece type fitted with HCN type canister. |
|---|---|

(This table is taken from DHEW (NIOSH) publication No. 77-108, Oct. 1976, **191**, p. 292. Criteria for a recommended standard – Occupational exposure to hydrogen cyanide and cyanide salts.)

## Occupational health

Because of the toxic nature of hydrogen cyanide and cyanide salts there is an absolute requirement by all staff to observe the strictest adherence to safety rules governing the ventilation of the workplace and the wearing of approved clothing and respiratory protection in areas containing cyanides. Medical surveillance and annual medical examinations should be available to all workers regularly handling or using cyanides. Small amounts of cyanide are detoxicated in the liver and excreted in the urine as the less toxic thiocyanate. As thiocyanates are found in some food and the blood of smokers, the level of thiocyanates in the urine cannot be used as an accurate measure of exposure to cyanide.

**First aid**

Cyanide is rapidly absorbed via the skin, lungs and the stomach into the blood. Cyanide rapidly passes to the cells and tissues combining with a vital respiratory enzyme cytochrome oxidase, preventing it from utilizing molecular oxygen in cellular respiration. The first aid treatment for cyanide poisoning is based on the principle of offering cyanide present in the blood another molecular site(s) which is as or more stable than the one present in the vital oxidase. Methaemoglobin and the cobalt chelating agent, cobalt EDTA both achieve this aim.

Amyl nitrite treatment is effective in mild poisoning because the volatile nitrite enters the body via the lungs and converts some haemoglobin in the blood to the methaemoglobin form. This readily forms a stable and inert cyanomethaemoglobin complex with cyanide in the blood. Quite high concentrations of methaemoglobin can be tolerated.

**The fatal dose for hydrogen cyanide is 50 mg and for cyanogen and cyanogen chloride is 100 mg. Death occurs in 1 to 15 minutes without treatment.**

**Speed and efficient treatment of a victim is vital to survival. Remove subject from the contaminated area using breathing apparatus. Call medical assistance at once.**

**Lay the patient down, take off any contaminated clothing to prevent further absorption of cyanide, and try and maintain warm and at rest. If the patient is breathing, break a capsule of amyl nitrite liquid onto a cloth or paper tissue and encourage the patient to breathe the vapour for 15 seconds. Repeat this procedure, if necessary, up to five times. If the patient is not breathing properly, give artificial respiration using an oxygen resuscitator (not mouth-to-mouth method).**

**In severe cases of poisoning, i.e. comatose patients not responding to amyl nitrite inhalation, intravenous infusion of 10 ml of 3% sodium nitrite solution given at the rate of 5 ml per minute followed by 50 ml of 25% sodium thiosulphate solution given at the same rate is required in addition to basic first aid. This administration should be given as soon as possible by a medically qualified person or paramedic.**

Cobalt EDTA treatment, as 50 ml of a 1.5% solution given intravenously, has been reported as very effective in cases of cyanide poisoning not responding to nitrite/thiosulphate treatment (Nagler et al. (1978))*. This treatment can lead to dangerous cardiovascular side-effects.

*Nagler, J., Prevoost, R.A. and Parizel, G. (1978). Hydrogen cyanide poisoning – Treatment with cobalt EDTA. *J. Occup. Med.*, **20**(6), 414–416.

# Hydrogen fluoride HF

**Exposure limits**
OES = 3 ppm (2.5 mg/m³)
These values relate to a Short-Term Exposure
Limit (10 minute reference period) only.

## Hydrofluoric acid

## Description

Hydrogen fluoride can exist in both liquid and vapour forms. **The vapour is a colourless, extremely toxic, corrosive gas that fumes in moist air. Both the liquid and the vapour should be treated with extreme caution.**

## Properties

The gas has a sharp irritating smell at low concentrations which can give good warning of its presence. It is very soluble in water and solutions of hydrofluoric acid are capable of etching or dissolving glass.

MW = 20,   SG = 0.99 (liquid),   MP = −83.1°C,   BP = 19.5°C,   VD = 1.27,
VP = 400 mmHg ($53.2 \times 10^3$ Pa).

## Detection methods

Leaks from equipment or cylinders can be detected by means of white fumes from an applied aqueous solution of ammonia.
Chemical reaction tubes.

## Precautions when using the gas

**Because of the extreme toxicity of both the liquid and vapour forms of hydrogen fluoride, great care is required in its storage, handling and use particularly when large quantities are involved.** All areas where the gas is used or stored must be adequately ventilated so that the concentration of vapour in the workplace is well below the Occupational Exposure Standard. All staff must be instructed in the hazardous nature of the substance and fully trained in the safety measures and personal protective equipment needed* when dealing with normal work, and in the event of accidental spillages of the liquid. In factories it is essential to provide adequate fast acting safety showers, drenches and eye wash fountains.

---

*The level of personal protection needed will depend on the scale of operations carried out. For small laboratory work, this would require neoprene apron, gloves and boots worn over a chemical resistant overall, and face and eye protection. For large industrial operations, a fully enclosed chemical suit with airline hood would be suitable. Staff handling this sort of dangerous material must not work alone.

## Occupational health

A vapour concentration of 50 ppm in air breathed for more than 30 minutes is fatal. Lower concentrations cause irritation to the mucous membranes and respiratory tract. The eyes are particularly sensitive to hydrogen fluoride vapour. Inhalation of low concentrations of the gas may not appear to have any effect but lung oedema may occur 12 to 24 hours following exposure.

---

**First aid**

Exposure to hydrogen fluoride should be considered a serious medical emergency.

The presence and/or advice from a medically qualified person is required as soon as possible.

Inhalation

The subject should be removed to fresh air and 100% oxygen administered, preferably by means of an oxygen resuscitator. Even with minor exposures, 100% oxygen should be given at intervals of up to 4 hours following exposure. **In severe exposures, the victim should receive oxygen therapy immediately, and then removed to hospital by ambulance without delay.** A member of staff should accompany the casualty to hospital to give details to the doctor of the type and duration of the exposure to assist in the further treatment.

Eye contact with vapour or liquid

**Without delay, irrigate the eyes and face with large volumes of cold tap water for at least 15 minutes. Get an assistant to call for an ambulance.** Repeat the irrigation treatment if necessary. The subject should be removed to hospital and then seen by an ophthalmic consultant.

Skin contact

**Immediately drench the affected area or whole body with a shower of cold water. Remove the victim's clothing while under the water shower. Continue cold water drenching for at least 10 minutes to remove as much fluoride from the skin. Get an assistant to seek medically qualified assistance as soon as possible.** Treat the affected area with iced 70% ethyl alcohol (ethanol) or an iced saturated solution of magnesium sulphate for 30 minutes. If medical aid is not available, continue the iced alcohol or magnesium sulphate treatment for two hours, and then apply a paste of magnesium oxide in glycerine. Alternatively, use a liberal application of 10% calcium gluconate gel to the affected area.

(Some medical practitioners prefer subcutaneous injections of 10% solution of calcium gluconate but the use of calcium gluconate gel has proved to be effective in treating liquid splashes and vapour burns.)

---

# Hydrogen selenide $H_2Se$

**Exposure limits**
OES = 0.05 ppm (0.2 mg/m³)
STEL = not quoted

## Description
Clourless, heavier than air, **highly toxic and flammable gas**, possessing a disagreeable penetrating odour.

## Properties
**All exposures to hydrogen selenide should be considered dangerous.** The gas decomposes at 16°C to give selenium and hydrogen. The selenium present in the resulting vapour is particularly toxic when inhaled.

## Detection methods
Not readily available.

## Precautions when using the gas
**Adequate safety and first aid arrangements must be available before any work is started with hydrogen selenide.**

Cylinders of the gas must be stored in the open air or in a well-ventilated exhaust protective cabinet. **Staff must not be allowed to work alone with this gas.**

## Occupational health
Slight exposure to the gas produces irritation of the eyes and the mucous membranes. Prolonged exposure to air containing <0.2 ppm of hydrogen selenide produces a garlic odour in the breath, nausea, dizziness and extreme fatigue.

The sudden inhalation of hydrogen selenide in the case of a laboratory accident involving a gas release can produce pulmonary oedema.

**First aid**

Inhalation

Remove victim from the contaminated area into fresh air, the rescuers wearing compressed air breathing apparatus. **Do not leave the casualty unattended.** If the patient's breathing become weak or stops, give artificial respiration preferably with an oxygen-supplemented resuscitator. **The casualty must be taken to hospital as soon as possible following the exposure.**

Skin contact

Casualties who have been contaminated with the gas or liquid should be immediately treated by thoroughly washing the affected part with a steady stream of cold tap water for at least 15 minutes, ensuring that as much of the hydrogen selenide as possible has been removed. If the skin has been burnt, cover with a dry dressing or clingfilm.

Eye contact

Thoroughly irrigate the eyes with a stream of cold tap water for at least 15 minutes. Cover the eye with a dry eye pad dressing **and refer the victim to hospital without delay for specialist treatment**.

# Hydrogen sulphide $H_2S$

**Exposure limits**
OES = 10 ppm (14 mg/m³)
STEL = 15 ppm (21 mg/m³)

### Description
Colourless, heavier than air gas which is **toxic, corrosive, if moist, and flammable.** It smells of rotten eggs. It is used industrially for the production of organic and inorganic sulphur compounds.

### Properties
The gas is soluble in water, ethyl alcohol, and oil. It burns in air to produce sulphur dioxide which is a suffocating toxic vapour. **Mixtures of hydrogen sulphide and air can form an explosive mixture. Contact with oxidizing agents may produce very violent reactions and spontaneous ignition.**

MW = 34,  D = 1.54,  MP = −85.5°C,  BP = −60.7°C,
VD = 1.19,  VP = 20 atmos. (2026 × 10³ Pa)
EL = 4.3–46%,  IT = 260°C.

### Detection methods
Moist lead acetate paper.
Chemical reaction tubes.
Portable electronic gas detector.

### Precautions when using the gas
**Staff should be warned of the dangers of working with hydrogen sulphide.** Laboratory work should always be carried out in a tested exhaust protective cabinet, so that the exhaust air is well below the OES of 10 ppm.

Industrial processes that produce hydrogen sulphide gas should be sited in a special well-ventilated area, the extracted air may need to be chemically treated with a gas scrubber to remove excess $H_2S$ gas.

In manufacturing or sewage treatment works, frequent measurements of the gas concentration in air must be made to ensure air pollution is being controlled.

### Occupational health
Workers who are involved in processes regularly using hydrogen sulphide should receive a pre-employment medical examination, and thereafter examined every 6 months.

Low concentrations of the gas are readily detected by smell, but because prolonged exposure to the gas dulls the sense of smell, this is a very unreliable method of hazard warning.

At concentrations of 10–500 ppm., the gas causes headaches which may last several hours, coughing and pain in the respiratory tract, dizziness, nausea and some eye irritation.

In moderate poisoning, exposures of 500–700 ppm, the victim will lose consciousness for several minutes and exhibit slate blue cyanosis until breathing restarts.

Inhalation of very high concentrations, >700 ppm, will produce respiratory arrest, anoxia and death.

Hydrogen sulphide enters the body via the respiratory system and it has a generalized cytotoxic effect by inhibiting the iron present in cytochrome oxidase, a vital enzyme required in cellular oxidation. Elimination of the sulphide occurs via the urine, faeces and the breath.

---

**First aid**

**Remove the victim to fresh air, wearing compressed air breathing apparatus.** Make sure the patient's airway is clear if breathing is weak, apply artificial respiration with supplementary oxygen if available. **Do not use mouth-to-mouth resuscitation.** The patient should be taken to hospital unless the exposure is trivial.

Skin or eye contact with the gas or a liquid containing hydrogen sulphide should be treated with a thorough irrigation of the affected area with a stream of cold tap water for at least 15 minutes. **Medically qualified advice must then be sought for follow-up treatment.**

# Inert gases

## Helium and other rare gases

## Nitrogen

**Exposure limits**
OES = None quoted
Ceiling concentration, the concentration of these gases must not lower the oxygen concentration in the air to less than 18% by volume.

### Description
Colourless, odourless, non-flammable and non toxic gases. Nitrogen is present as 78.5% in the air.

### Properties
Nitrogen is very slightly soluble in water and ethyl alcohol. The other rare gases are chemically inert.

| Name | AW | SG | MP°C | BP°C |
|------|------|------|--------|--------|
| nitrogen | 14.0 | 1.25 | −209.8 | −195.8 |
| helium | 4.0 | 0.18 | −272.2 | −268.9 |
| neon | 20.2 | 0.90 | −248.6 | −245.9 |
| argon | 39.9 | 1.78 | −189.2 | −185.7 |
| krypton | 83.8 | 3.74 | −156.6 | −153.3 |
| xenon | 131.3 | 5.89 | −111.9 | −107.1 |
| radon | considered under a separate heading | | | |

Note: Helium is often substituted for nitrogen in compressed air supplied to deep sea divers. Helium reduces the tendency for nitrogen under pressure to dissolve in body lipids particularly those present in the nervous system. Rapid decompression from compressed air releases dissolved nitrogen as gas bubbles (diver's bends) which can damage the nervous system.

### Detection methods
Chemical methods of detection of these gases are not possible as they are inert at room temperature. Burning magnesium will react with nitrogen to form magnesium nitride. The rare gases are chemically inert.

### Precautions when using the gas
Nitrogen or rare gases should not be used in confined spaces since a substantial leakage from a cylinder or gas line could produce dangerously low levels of oxygen by displacement.

## Occupational health

High concentrations of any of these gases will produce a breathing mixture devoid of an adequate percentage of oxygen. This will rapidly produce anoxia and unconsciousness.

---

**First aid**

Inhalation

If unconscious or drowsy, the subject should be removed to an uncontaminated area and allowed to breathe fresh air. Note: If the patient is unconscious the atmosphere is probably dangerous, therefore **self-contained breathing apparatus must be worn (not a gas mask) by the rescuer(s)**. Remove casualty to a safe area and administer oxygen-enriched air or pure oxygen. Artificial respiration may be necessary using an oxygen-enriched resuscitator such as an Ambubag. **Casualties that have been unconscious must be seen by medically qualified staff as soon as possible after the incident.**

Skin contact (with liquefied gas)

Liquefied gases can produce very serious cold burns when in contact with skin. The affected area should be immersed in tepid water or cold water to reduce the cooling action of the liquefied gas and reduce tissue damage. As soon as possible the victim should be removed to hospital for further specialized treatment. Recovery from a severe liquefied gas burn can be very painful and the casualty must not be left unattended.

Eye contact (with liquefied gas)
THIS IS A MEDICAL EMERGENCY.

Immediately irrigate the eye with cold tap water for 15 minutes to remove the liquefied gas and warm up the affected tissues to reduce damage. Cover the eye(s) with a loose dry dressing **and arrange for the transfer of the casualty to hospital without delay for specialist medical treatment.**

---

# Iodine

$I_2$

**Exposure limits**
No OES (TWA) value quoted
STEL = 0.1 ppm (1 mg/m³)

### Description
Iodine is a blue-black solid with a metallic lustre possessing a characteristic odour and sharp metal taste. Iodine solid and vapour is a lung and skin irritant, and can easily be concentrated in the thyroid gland causing metabolic disturbances.

### Properties
Iodine is the least chemically reactive than the other halogens. The element is slightly soluble in water but readily soluble in organic liquids, and forms many iodine interhalogen compounds and iodides.

$MW = 253.8$,  $SG = 4.93$,  $MP = 113.5°C$,  $BP = 184.3°C$,
$VP = 1 \, mmHg$ ($0.13 \times 10^3$ Pa) at 38.7°C.

### Industrial uses
Because iodine and iodides are expensive, many industrial processes that once used iodine have been superseded by other reagents. Sodium and potassium iodides are produced for use in the chemical industry, in the production of silver iodide for photographic film, and potassium iodide as an additive to 'iodized salt' which contains 0.023% potassium iodide to prevent iodine deficiency in humans (goitre) and animals.

### Detection methods
Iodine present in solution can be easily detected by adding a few drops of the solution to be tested into a dilute starch solution. The presence of iodine is seen by an intense blue coloration (sensitivity 1 part iodine in 5 million of water).

Iodine present in air can be determined by passing a known volume of air through a complexing solution such as 3% potassium iodide, toluene, or ortho-toluidine and determining the concentration of the complex by an ultraviolet spectrophotometer.

### Precautions when using the substance
**The STEL value indicates that iodine vapour is extremely irritant to the respiratory system, and prolonged exposure is not permissible.** Iodine will also produce unpleasant eye and skin irritations which can be similar to corrosive chemical burns leading to slow-healing ulcers. It is essential that all work involving iodine in the solid or vapour form must be carried out inside a closed system or protective cabinet.

If any contact with the substance is likely, chemical protective clothing with hand and foot protection is necessary. Full face gas masks or airline hoods must ensure respiratory and eye protection for the workers.

## Occupational health

Staff involved with regular work involving iodine or its compounds should receive a special pre-employment medical examination. Staff who show signs of bronchitis, emphysema, or lung or thyroid disorders should not be employed. Great care should be exercised in the maintenance of good hygiene standards in the workplace, and in the choice of personal protective equipment. The exhaust ventilation equipment used in areas handling iodine should be fitted with air scrubbers or filters to remove iodine from the exhaust air before discharge into the environment.

---

**First aid**

Inhalation
Remove the casualty from the contaminated area into fresh air, the rescuers wearing protective clothing and full face gas masks fitted with an iodine absorbent canister. The victim should be assessed as to the severity of the exposure. If unconscious, carry out resuscitation using an oxygen-enriched air resuscitator. If conscious, place at rest in a slightly raised position to reduce the possibility of lung oedema, give oxygen if required, and keep the patient warm. **Transfer to hospital without delay for specialist medical attention.** Inform the medical staff of the type and, if possible, duration of the exposure.

Ingestion
Ingestion is rare since it usually causes vomiting by its own irritant action on the stomach. Following vomiting rinse out the patient's mouth with cold water and then encourage them to drink a glassful of milk. Keep the patient warm and at rest **and refer them to hospital without delay.**

Skin contact
Skin lesions from iodine resemble thermal burns but with brown stains. Any iodine contact with the skin should be immediately treated by irrigation of the affected area with cold tap water for at least 10 minutes. Liquid paraffin, soaked gauze should be laid over the affected area and the casualty removed to hospital for specialist treatment.

Eye contact
**Solution or vapour contact with the eye is extremely serious.** Immediately irrigate the eye and surrounding tissues with large volumes of cold tap water for at least 15 minutes. Cover the eye with a dry eye pad, keep the patient warm and at rest **and then transfer the patient without delay to the nearest hospital for further specialist eye treatment.**

---

# Iron and compounds     Fe

**Exposure limits**

Iron oxide fume (as Fe)  OES (TWA) = 5 mg/m³
STEL = 10 mg/m³
Iron salts (as Fe)  OES (TWA) = 1 mg/m³
STEL = 2 mg/m³

## Description
Iron as a metal exists as a tough, malleable, steel grey metal used in the construction and manufacturing industry in vast quantities. Iron containing ores are abundant in nature and consist of oxides, $Fe_3O_4$ (magnetite) and $Fe_2O_3$ and $FeCO_3$ (ferrous carbonate). Iron is prepared from iron ores by reduction with carbon in a blast furnace, producing pig iron which is used in steel manufacture.

## Properties
Metallic iron is soluble in mineral acids, but insoluble in water. The surface of iron easily forms an oxide coating as a result of oxidation in the presence of water.

AW = 55.85,  SG = 7.86,  MP = 1535°C,  BP = 2750°C.

Iron salts consist of ferrous ($Fe2^+$) and ferric ($Fe3^+$) compounds. Ferric compounds are the most stable.

## Industrial uses
The major industrial use of iron is in the manufacture of steel and specialized iron products.

Many iron compounds are used in industry. Iron acetate, chloride and nitrate are used as fabric mordants in the dyeing industry. Ferric oxide, chromate and ferrocyanides are used as pigments. Ferric hydroxide is used as a precipitating reagent in the purification of water.

## Detection methods
Colorimetric methods can be used to detect trace quantities of iron present in solutions. There are many complexing agents that form coloured products with iron salts notably cupferrate. Unfortunately many colorimetric methods are affected by interference from other metals present.

Flame and flameless atomic absorption spectrographic methods are the most reliable for iron determinations particularly if the iron is present in a biological matrix.

## Precautions when using the substance
Iron is particularly inert when handled. Processes that produce quantities of fine dust either as fine iron dust or oxide can produce abnormal changes in the lungs

described as iron siderosis, presenting as a form of pneumoconiosis. This condition can give rise to a chronic bronchitis. To prevent this, workers should have a working environment with a low dust level (i.e. $<10\,mg/m^3$ air), or if this is not practicable, wear respiratory protective equipment to remove dust from the respired air. In the case of welders working with iron or steel, suitable dust respirators and exhaust protective cabinets should be provided. (Note: In the last few years, safety equipment manufacturers have produced a range of ventilated safety helmets that incorporate a welding visor which has a stream of filtered air passing across the head and face of the wearer. The safety helmets are light to wear, and keep the face and head of the wearer cool during the welding operations.)

## Occupational health

Acute poisoning from iron is unknown. Poisoning from iron salts is rare but can be fatal. It usually occurs in cases of self-administration of soluble iron salts, e.g. ferrous sulphate. (Note: ferrous $Fe^{2+}$ salts are more easily absorbed via the gastrointestinal tract than ferric $Fe3+$ salts.)

Although iron is required as an essential trace element in humans for the synthesis of haemoglobin and vital cellular enzymes, only 1.5 mg of ferrous iron salts are needed per day to maintain normal metabolism. If large amounts of soluble ferrous salts are ingested, they inhibit the normal mucosal block on the uptake of iron, and allow large quantities of iron to be absorbed. This could lead to acute poisoning requiring specialized hospital treatment. The treatment of choice is to give desferroxamine by mouth to chelate the iron, which can then be excreted.

Dimercaptopropanol (BAL) is contraindicated because the iron–BAL complex is toxic.

**First aid**

### Inhalation

The person exposed to iron or iron oxide dust should be removed from the contaminated area to fresh air. They should be allowed to rest and kept warm, and examined by a medical practitioner for signs of siderosis and/or lung function impairment.

### Ingestion

**Ingestion of soluble iron salts can be fatal. Where there is a reasonable suspicion that a significant quantity of iron salts (e.g. ferrous sulphate) has been taken, this must be treated as a medical emergency.**

If the victim is conscious, give a glass of milk to drink to buffer the corrosive effect of the iron salts and to delay passage of the poison into the duodenum. Try to make the victim vomit and then give more milk. If milk is not available give plenty of water to drink The patient at this stage may complain of stomach pains and nausea.

Keep the patient warm and at rest **and arrange for their removal to hospital without delay**. If the patient looks pale or grey in colour with rapid respiration and pulse rate, give oxygen or oxygen-enriched air.

Inform the medical staff at the hospital of the nature, if possible, name of the substance ingested and the quantity, if known.

### Skin contact

Wash the contaminated area with a stream of cold tap water for at least 10 minutes, to remove as much of the iron salts as possible. If the skin appears normal, cover with a dry dressing. If the skin appears red and/or inflamed, treat as a chemical burn, cover with a dry dressing, and refer the patient to hospital for further specialized treatment.

### Eye contact

**Any exposure of the eyes to iron salts is a serious medical matter.** Immediately irrigate the eye and surrounding tissues with a stream of cold tap water for at least 15 minutes making sure that the whole area of the eye and the surrounding orbit is well washed with water. Cover the eye with a dry eye dressing **and immediately refer the patient to hospital for specialist medical treatment.**

# Lead (Pb) and compounds

## Exposure limits

For lead and lead compounds (except lead tetraethyl)   OES (TWA) = 0.15 mg/m³ in air
For lead tetraethyl                                     OES (TWA) = 0.10 mg/m³ in air
Urinary lead concentration <65 µg/100 ml, blood lead concentration <80 µg/100 ml with
a zinc protoporphyrin IX concentration of <3 µg/g haemoglobin.

## Description

Lead is a heavy blue-grey metal which oxidizes in moist air. It is very soft and malleable, easily forms lead alloys with arsenic, tin and bismuth is used and in the manufacture of brass, bronze and certain steels. Lead compounds are used widely in industry. They exist in two series of compounds: the stable $Pb^{2+}$ and the less stable $Pb^{4+}$. Lead is also amphoteric forming both $Pb^{2+}$ and $PbO_2{}^{2-}$ salts. **Lead compounds, particularly the soluble salts and lead alkyls, are cumulative poisons and can damage major organs of the body including the nervous system.**

## Properties

**Metallic lead** is slightly soluble in cold water, but much more soluble in hot water, low in calcium content and with high acidity. This latter property has led to the banning of lead pipework in water supply systems, and the substitution of copper or plastic.

AW = 207.2,   SG = 11.3,   MP = 327°C,   BP = 1525°C.

**Lead monoxide PbO**, a yellow powder formed by heating lead in air.

**Lead hydroxide Pb(OH)$_2$**, a white gelatinous precipitate formed by the action of a strong alkali on a lead salt solution.

**Lead chloride PbCl$_2$**, a white precipitate formed by the action of hydrochloric acid on a lead salt. It is used as the starting material for the preparation of crystalline lead containing complex salts.

**Lead carbonate PbCO$_3$** is a white crystalline precipitate formed when a solution of an alkali carbonate is added to a lead salt in the cold. The basic carbonate, $2PbCO_3$, $Pb(OH)_2$, is known as white lead.

**Lead acetate Pb(CH$_3$CO$_2$)$_2$, 3H$_2$O** is prepared by dissolving lead oxide in hot dilute acetic acid. Known as 'sugar of lead' it is extremely poisonous.

**Lead sulphate PbSO$_4$** is formed as a dense white precipitate by the action of sulphuric acid on a lead salt solution.

**Lead arsenate PbHAsO₄** is produced by reacting lead oxide with arsenous acid in the presence of nitric acid. **It is an extremely toxic compound, particularly if heated, when toxic fumes are emitted.** The compound contains 23% arsenic and 63% lead.

MW = 327, SG = 5.79, MP = 270°C.

## WARNING
**Lead azide, lead fulminate and lead perchlorate are all compounds which are extremely unstable lead salts and can detonate or explode without warning.** Their metastable nature may be due to strained molecular structure in the crystal.

## Industrial uses
Metallic lead is used as protective sheathing for cables, as screening material against gamma radiation, and as jointing material in the automobile industry. Lead is used extensively in the preparation of metal alloys. Lead and lead oxides are used in the manufacture of lead acid batteries. Lead salts are used widely in the paints, pigments, lacquers and varnishes, in printing and publishing, and in the glazes used in the pottery industry. Some of the very toxic lead salts are used as pesticides.

## Detection methods
Lead can be determined easily by a number of standard analytical methods.

Colorimetric methods involving the use of the production of coloured lead complexes such as lead dithizone, benzoyl acetonate, cupferron or beta-naphthyl thiocarbazone.

DC arc and X-ray spectroscopy have been successfully used for the determination of lead in blood and tissues.

Polarography, using pretreatment of the sample by co-precipitating the lead present in solution with calcium oxalate, can offer a high precision analytical method in the microgram range.

Atomic absorption spectrophotometry has been used to measure lead in blood and tissues. Recently highly sensitive methods have been devised to determine lead concentration in 20 µl of blood using a micro cup or carbon furnace. Tissue samples would need acid digestion followed by a chelation extraction procedure prior to using the carbon furnace technique.

Air samples containing lead can be estimated by passing a known volume of air through a gas extraction bottle containing a liquid which will remove any airborne lead. The lead concentration present in the liquid could then be estimated by one of the methods outlined above.

## Precautions when using lead or a lead compound
Lead and its compounds are toxic and they can easily enter the body by inhalation of dust or fumes, or ingestion or be absorbed through the skin particularly if the lead is present in an oil- or grease-based product. Lead entering the body tends to act as a cumulative poison to the major tissues in the body particularly the blood, liver, kidneys and the nervous system.

The dangers of occupational lead exposures have been known for many years and have led to strict Codes of Practice designed to protect the worker and reduce exposure from lead to a minimum, and below the clinically significant level. In the UK the Approved Code of Practice – Control of lead at work – (HMSO, London, 1980) covers some twenty previous Regulations established over 50 years.

The new Code of Practice requires that any workplace using or fabricating lead must:

1. carry out a risk assessment of work which might expose workers to the effects of lead,
2. provide adequate control measures to contain lead in the plant, any processes and in the storage of materials,
3. ensure the control measures are used and properly maintained,
4. provide, where necessary, protective clothing and respiratory protection against dusts and fumes,
5. provide adequate means to prevent the spread of contamination by lead in the workplace and in the environment,
6. provide regular monitoring of the air inside and outside the workplace,
7. provide full medical surveillance and biochemical tests related to exposure of all workers involved with lead, and the keeping of medical and analytical records of all staff surveyed.
8. provide full information, instruction and training given to all staff using lead including the keeping of all records of such procedures.

The Code set down strict limits for exposure to lead by workers.
**The OES for the lead in air concentration must not exceed 150 µg/cubic metre of air.**

Workers with a blood lead concentration of >80 µg/100 ml (women of reproductive age must not exceed 40 µg/100 ml of blood) must be certified unfit for further work with lead, and not be allowed to continue such work.

## Occupational health
Inorganic lead ingestion produces a number of important biochemical changes in the body. Lead is readily absorbed from the gut or the lungs and is transported throughout the body by the red cells. The presence of significant levels of lead in blood gives rise to inhibition of the activity of some vital enzyme systems notably delta-aminolaevulinic acid dehydrase which causes an increase of this aminolaevulinic acid in the blood and the urine, leading to its transport across the blood–brain barrier. In children this is thought to be responsible for lead-induced hyperactivity. Lead circulating in the body also affects the enzyme systems responsible for the synthesis of haem from protoporphyrin IX and iron. The result of this inhibition is the formation of zinc protoporphyrin IX instead of haem. These changes give rise to hypochromic anaemia and degenerate erythrocyte synthesis.

Early manifestations of lead intoxication are usually seen in severe disturbances of the gastrointestinal tract leading to persistent diarrhoea, and peripheral and central nervous system dysfunction. Long-term exposure to lead may cause paralysis of parts of the peripheral nervous system, particularly in the radial nerve ('wrist drop'). Workers who are exposed to uncontrolled lead contaminated environments can develop lead colic in the months following exposure. This condition gives rise to severe stomach cramps, constipation and vomiting, and can be misdiagnosed as appendicitis. Severe long-term exposure can produce lead encephalopathy, where the high lead concentrations in the blood pass through the blood–brain barrier and damage the brain cells producing epileptiform fits, coma and death. Fortunately this condition is now rarely seen because of stringent occupational hygiene working standards.

**First aid**

### Inhalation

Workers who have inhaled significant quantities of lead dust or fumes containing lead should be kept warm and at rest. They should be removed to hospital by ambulance and their exposure explained to the hospital medical specialist who will decided on any further treatment.

### Ingestion

Ingestion of lead dust is less of a problem that inhalation. If corrosive lead salts are ingested, they can be rapidly absorbed into the blood-stream leading to lead poisoning. The victim should be given milk to drink to delay the passage of the lead salts from the stomach, and then encouraged to vomit to empty the stomach. The patient should then be removed to hospital for medical assessment to see if lead chemotherapy is required. This consists of intravenous administration of calcium disodium versenate in glucose saline and monitoring the excretion of the lead versenate complex in the urine, until the level is within the normal range. Versene therapy does have unpleasant side-effects if the treatment is given incorrectly. The long-term effects of lead intoxication can be controlled by oral administration of 20–40 mg/day of D-penicillamine. The lead penicillamine complex is water soluble and readily excreted in the urine.

### Skin contact

Any spillage of lead or lead containing salts must be removed without delay using luke warm water and soap. If the skin exposure is on the hands only, this washing technique can be used to remove lead. In some workshops where splashing onto protective clothing occurs regularly with lead salts, versene containing soap is provided to assist in the removal of lead. If the accident involves a spillage of lead salts over the whole body, the victim must be taken to a shower and washed with luke warm water and soap for 10 minutes to remove all traces of lead. The casualty should be referred to a medically qualified occupational health physician for further advice.

### Eye contact

Any accident involving lead particles or lead salts in the eye must be treated carefully. The patient should be treated by irrigating the eye with a steady stream of cold water to remove any lead salts or if possible flush out any particles from the surface or orbit of the eye. The affected eye(s) should be covered with a loose dry dressing **and the patient referred without delay to the nearest eye specialist at the Accident and Emergency department**.

# Lead alkyl compounds

### Exposure limits
For tetramethyl lead
OES (TWA) = 0.10 mg/m$^3$ in air measured as Pb
For tetraethyl lead
OES (TWA) = 0.10 mg/m$^3$ in air measured as Pb

EOHS give TLV and STEL values as follows:
Tetramethyl lead TLV (ACGIH) = 0.15 mg/m$^3$ for skin contact
STEL (ACGIH) = 0.5 mg/m$^3$ for skin contact
Tetraethyl lead TLV (ACGIH) = 0.10 mg/m$^3$ for skin contact
STEL (ACGIH) = 0.30 mg/m$^3$ for skin contact
all these values measured as lead.

Tetramethyl lead    $(CH_3)_4Pb$

Tetraethyl lead    $(C_2H_5)_4Pb$

## Description
Many lead alkyl compounds have been synthesized and isolated by organic chemists but the most significant industrially are tetramethyl and tetraethyl lead. These compounds are used as an additive in the formulation of petroleum fuel for use in high compression engines.

## Properties
### Tetramethyl lead
Tetramethyl lead is a colourless oily liquid with a characteristic pleasant smell. It is insoluble in water, slightly soluble in benzene, ethanol and diethyl ether:

MW = 267.3,   SG = 1.99,   MP = –27.5°C,   BP = 110°C,   VD = 9.2,
VP = 22 mmHg (2.92 × 10$^3$ Pa) at 25°C,   FP = 37.7°C,   EL = 1.8% (lower limit).

### Tetraethyl lead
Tetraethyl lead is a colourless oily liquid with a characteristic pleasant smell. It is insoluble in water but soluble in benzene, ethanol and diethylether.

MW = 323,   SG = 1.66,   MP = –136.8°C,   BP = 200°C (with decomposition),
VP = 1 mmHg (0.13 × 10$^3$ Pa) at 38.4°C,   FP = 93.3°C.

## Industrial uses
Lead alkyls particularly tetraethyl lead have been used for many years as a pre-ignition suppression additive to petrol. The organic component of the molecule increases the effective 'octane number' of the fuel and the presence of lead can act as a lubricant in the engine cylinder linings. With the development of the newer 'lean burn' engines fitted with hardened valves, the need for lead alkyl additives is

reduced. Legal limits of 1.12 g/litre have been set by many countries for this additive to minimize lead emissions in exhaust gases.

## Detection methods
Lead can be determined easily by a number of analytical methods.

Colorimetric methods involving the use of the production of coloured lead complexes such as lead dithizone, benzoylacetonate, cupferron or beta-naphthyl thiocarbazone.

DC arc and X-ray spectroscopy have been successfully used for the determination of lead in blood and tissues.

Polarography using pretreatment of the sample by co-precipitating the lead present in solution with calcium oxalate, can offer a high precision analytical method in the microgram range.

Atomic absorption spectrophotometry has been used to measure lead in blood and tissues. Recently, highly sensitive methods have been devised to determine lead concentrations in 20 microlitres of blood using a micro cup or carbon furnace. Tissue samples would need acid digestion followed by a chelation extraction procedure prior to using the carbon furnace technique.

Very low levels of lead alkyls can be measured using an atomic absorption spectrophotometer coupled to a gas chromatograph. Lead alkyls at a concentration level of 0.1 nanograms can be measured.

Air samples containing lead can be estimated by passing a known volume of air through a gas extraction bottle containing a liquid which will remove any airborne lead. The lead concentration present in the liquid could then be estimated by one of the methods outlined above.

## Precautions when using the substance
**All lead alkyl compounds are particularly toxic if absorbed into the body through the skin or by inhalation of vapour or aerosol.** It is essential that industrial workers handling undiluted lead alkyls should wear suitable personal protective clothing, preferably a chemical resistant suit with neoprene gloves or gauntlets, and have availability of canister type gas masks or fresh air positive pressure hoods in the event of a vapour leakage. Good personal and workshop hygiene must be observed to reduce the possibility of contamination with any alkyl liquid. Most workplaces ensure a low level of alkyl vapour by operating exhaust protective ventilation at vapour emission points in the process, and very good general ventilation throughout the workplace.

## Occupational health
**A very careful control of the working environment and the exposed staff is required to prevent clinical exposures to lead alkyls.** Pre-employment and regular medical examinations are essential to protect staff from the effects of working with lead. Regular screening of blood and urine for lead levels in process workers and the concentration of lead alkyls in the air must be carried to ensure the health and safety measures are effective. EOHS suggests that persons suffering from mental disorders, alcoholism or hypertension should be excluded from working in potentially hazardous parts of the process. Urine levels exceeding 150 µg Pb/litre indicate an unacceptable amount of lead alkyl absorption. Urine levels of >180 µg Pb/litre require the worker to be removed from exposure and a careful investigation carried out to discover the nature and extent of it.

Reports of cases of tetraethyl lead poisoning indicate that fatalities usually occur from exhaustion. Even in severe cases, recovery is possible if the patient is given prolonged supportive therapy involving carefully controlled water and electrolyte balance and nutrition in association with long acting barbiturate treatment. These patients need careful and constant attendance due to a proneness to lapse into irrational behaviour.

---

**First Aid**

Inhalation

Workers who have inhaled a significant quantity of lead alkyl vapour should be removed from the contaminated area into fresh air. They should be kept warm and at rest. As soon as possible the victim should be removed to hospital by ambulance and the type of exposure explained to the hospital medical specialist who will decide on further treatment.

Ingestion

This is extremely rare. If liquid containing lead tetraethyl is accidentally drunk, the patient should be given a glass of milk to drink to delay the passage of any lead from the stomach, and encouraged to vomit to empty the contents of the stomach. The victim must then be removed to hospital for medical assessment for possible further treatment.

Skin contact

Any spillage of lead alkyl onto the skin must be removed without delay using luke warm water and soap. If the accident involves a spillage onto the whole body, the victim should be taken to a shower all clothing removed and the whole body washed very thoroughly with warm water and soap to remove all traces of the lead alkyl. The casualty should then be referred to a medically qualified occupational health physician for further advice.

Eye contact

Any accidents involving lead alkyls in the eye must be treated carefully. The patient should be treated by irrigating the eye with a stream of cold tap water, for at least 10 minutes, to remove as much lead alkyl contamination as possible. The affected eye(s) should be covered with a loose dry dressing **and the victim referred without delay to the nearest eye specialist at the Accident and Emergency department.**

---

# Lead arsenate $PbHAsO_4$

**Exposure limits**
OES (TWA) = 0.01 mg/m$^3$

## Description
A heavy white powder containing 63% lead and 23% arsenic. On heating it gives off toxic fumes containing lead and arsenic. **The powder is extremely toxic to humans if inhaled or ingested. The symptoms are of arsenic and lead poisoning.**

## Properties
Lead arsenate is insoluble in water, but soluble in dilute nitric acid and caustic alkalis. The powder has been used as a powerful insecticide, but has now been largely superseded by organophosphorus compounds.

MW = 327,  SG = 5.79,  MP = 270°C with decomposition.

## Industrial uses
The main use of the chemical is in the preparation of pesticides to combat sap-sucking insects such as aphids.

## Detection methods
No specific methods are available for the measurement of lead arsenate in the air. The presence of both lead and arsenic can be shown by chemical analysis.

## Precautions when using the substance
**Because of the extreme toxicity of this material, great care must be taken when it is manufactured, or the powder form is handled or used.** Persons working in environments using lead arsenate must wear protective clothing and full respiratory protective equipment to prevent any dust entering the respiratory tract. Some scale processes must be carried out inside an exhaust protective cabinet. Full washing and changing facilities must be provided for all workers handling or using the chemical. Workers must be required to remove their contaminated clothing and equipment, and wash or use a hot shower before changing into clean clothes at the end of work shifts. Without adequate washing, significant amounts of lead arsenate can be retained in the folds of the skin and ultimately lead to poisoning.

## Occupational health
**A full occupational health protocol must be approved and operational before large-scale operations are carried out using lead arsenate.** Staff must be regularly screened for the early symptoms of exposure to arsenic and lead. Measurement of

inorganic arsenic levels in urine, and blood lead and zinc protoporphyrin IX at regular intervals will indicate any unacceptable exposure.

---

**First aid**

Inhalation

**Lead arsenate dust is extremely toxic.** The patient should be kept warm and at rest preferably on a stretcher and removed to hospital without delay for specialist medical treatment. If the patient shows any respiratory distress, give oxygen or oxygen-enriched air using a face mask. **Do not give mouth-to-mouth resuscitation; if no oxygen mask is available, use a plastic airway.**

Ingestion

**Lead arsenate will be readily absorbed through the stomach wall particularly if the contents of the stomach are acid.** Immediately give the victim a glass of milk to reduce the acidity by buffering, and to reduce the rate of emptying of the stomach into the duodenum. If possible make the victim sick to remove any remaining lead arsenate powder from the stomach. **Take the victim to hospital without delay explaining to the medical staff the name of the substance ingested and the approximate amount, if known.**

Skin contact

Remove as much of the lead arsenate powder from the skin by washing the skin surface with cold water for ten minutes followed by careful washing with warm water and soap. The casualty should be seen by a medically qualified doctor for assessment.

Eye contact

Immediately irrigate the surface of the eye with a stream of cold tap water, continuing this treatment for at least 10 minutes, making sure that the surface and orbit of the eye are thoroughly washed. Cover the eye with a loose dry dressing **and refer the patient as soon as possible to an eye specialist at the nearest Accident and Emergency department.**

---

# Manganese Mn

**Exposure limits**

| | |
|---|---|
| Manganese compounds | OES (Ceiling value) = 5 mg/m³ |
| Manganese (as Mn fume) | OES (TWA) = 1 mg/m³ |
| | STEL = 3 mg/m³ |
| MMT (methyl manganese tricarbonyl) | OES (TWA) = 0.2 mg/m³ |
| | STEL = 0.6 mg/m³ (skin) |

## Description
Manganese metal exists as a grey-white brittle material, but the majority of manganese exists in soil as the stable oxide, manganese dioxide (pyrolusite).

## Properties
Manganese can exist in eight states of oxidation, the most important are $2^+$, $3^+$ and $7^+$. This means that a whole range of manganese containing compounds are possible, in the form of metallic manganese ions, and as complex anions such as manganates and permanganates.

AW = 54.9,   SG = 7.2,   MP = 1244°C,   BP = 1962°C.

## Industrial uses
Manganese metal is used in the production of steel as an additive to reduce sulphur and oxygen impurities. Manganese is also used to produce special alloys with copper, zinc, and aluminium to increase hardness. Manganese dioxide is used in the manufacture of dry cell batteries. Manganese salts are used in the dyeing and tanning industry, and as a mineral supplement in fertilizers. An organomanganese compound, methylcyclopentadienyl manganese tricarbonyl (MMT) is used extensively as a smoke retardant in fuel oil, and as a supplement to tetraethyl lead to prevent pre-ignition of fuel in petrol engines.

## Detection methods
Colorimetric methods involving the formation of highly coloured manganese complexes with formaldoxime have been used to determine its concentration in biological material. Coloured complexes are also formed with o-tolidine, 2-nitroso-1-naphthol and salicylaldoximate. Chelation of manganese with sodium diethyldithiocarbamate or its derivatives, followed by extraction of the complex into the solvent methylisobutyl ketone, has enabled manganese to be determined by both flame and flameless atomic absorption spectroscopy. DC arc spectroscopy has been used by some laboratories for the determination of the manganese content of blood, urine and hair in cases of poisoning.

## Precautions when using the substance
Although manganese is an essential trace element in the metabolism of animals and humans, the amount required for normal growth and development is small. A well-balanced diet contains about 3–7 mg of manganese per day.

Handling or being exposed to manganese containing dusts and fumes in the workplace can give rise to manganese intoxication. The main route of manganese entry into the body is via the lungs as fine dust or manganese containing fumes. The risk of manganese exposure from skin or gastrointestinal uptake is small. Therefore, to prevent occupational exposure to manganese, all work should be carried out in a dust-suppressed environment, using water sprays or dust extraction cabinets, and if necessary, e.g. in the mining of manganese ores, workers must wear adequate personal respiratory protection.

## Occupational health

Manganese in the form of fine dust or metal fume enters the lungs and is absorbed into the blood via the alveoli. The manganese probably forms a loose complex with the blood albumin protein fraction and is transported to the liver, and is then excreted into the gut via the bile. If the exposure to manganese is excessive, deposition of manganese can also occur in the kidneys, small intestine, endocrine glands and the bones. High blood manganese levels encourage the uptake of the metal into pigmented areas of the body such as the skin, hair and eyes, and the brain. **Deposition of manganese in the brain is a serious medical problem and irreversible.** The metal is thought to affect the working of the basal ganglia producing neurological symptoms of rigidity and walking spasticity very similar to patients with Parkinson's disease. Manganese poisoning has been treated in the early stages by using orally administered chelating agents such as calcium disodium versenate, which readily forms the stable manganese versenate complex which is removed by the kidneys into the urine. Some medical specialists have treated manganese toxic patients with large oral doses (up to 12 g/day) of L-dihydroxyphenylalanine (L-DOPA), a treatment originally given to some Parkinson's cases to replace the L-DOPA unable to be synthesized in the basal ganglia of their brains.

Exposure to manganese can be estimated by measuring the concentration of manganese in the urine. The urinary concentration should not exceed 50 mg/litre. Levels above this value indicate excessive exposure.

**First aid**

## Inhalation

Low or moderate levels of manganese dust or fume inhalation usually produce mild or no obvious symptoms. Exposed workers should be removed from the source of contamination and checked for any abnormal medical signs. A medical doctor should be contacted if there are any problems. Most cases recover completely in 24–48 hours following exposure.

Acute exposure can result in a condition known as metal fume fever. The patient has chest pain and difficulty in breathing. He/she may complain of an influenza like fever with sweating and shivering. The subject must be removed from the contaminated area into fresh air, rested and given oxygen if required. They should not be left alone until normal breathing is restored. The patient should be seen by a medically qualified person before returning home. Full recovery from these effects occurs in 1–2 days following exposure.

Note: Onset of metal fume fever symptoms can be delayed for some hours after exposure. In cases of fever occurring at night following a day work shift, medical staff should be informed of the patient's occupation.

## Skin contact

Any manganese salts, dust or powders should be removed from the surface of the skin using a stream of cold water and gently rubbing the surface of the skin with a soap solution, if the skin is unbroken. If lacerations are present, cold water only should be used and attempts made to prevent manganese from entering the wounds. Victims with wounds must be referred to hospital for further treatment.

## Eye contact

Manganese salts or dusts tend to attach themselves to the cornea of the eye causing intense irritation and pain. It is important to try and remove the foreign material by cold water irrigation of the eye using a slow stream of cold tap water for at least 15 minutes. Following this treatment the eye should be covered with a loose dressing **and the casualty referred to the nearest Accident and Emergency department for specialist treatment**.

# Mercury and inorganic compounds    Hg

**Exposure limits**
Mercury and inorganic compounds    OES (TWA) = 0.05 mg/m³
STEL = 0.15 mg/m³
New Regulations (1996) have reduced the OES level to 0.025 mg/m³

## Description
Mercury exists as a silvery white liquid metal at room temperature. Inorganic mercury compounds are poisonous and exist as $Hg1^+$ and $Hg2^+$ forms. The $Hg2^+$ salts are more soluble and therefore more toxic.

## Properties
Mercury metal is soluble in dilute nitric acid but insoluble in water. It readily combines with sulphur and halogens, and forms amalgams with most metals. This property has been used for many years to provide very durable mercury silver amalgam restorations in teeth.

Mercury $1^+$ compounds except the fluoride and nitrate are all sparingly soluble in water and not easily hydrolysed. Mercurous chloride (calomel), originally used as a laxative, is still used in some pharmaceutical preparations. Mercury $2^+$ compounds are very much more toxic because of their greater solubility particularly in hot solutions. Mercuric chloride (corrosive sublimate) is particularly toxic and an effective poison both to bacteria, animals and humans. Mercuric nitrate used originally in processing of felt hats is also very toxic if ingested or inhaled (mad hatter's disease).

AW = 200.7,   SG = 13.59,   MP = −38.9°C,   BP = 356.6°,   VP = 0.012 mmHg (1.5 Pa) at 20°C but 1 mmHg ($0.013 \times 10^3$ Pa) at 26°C.

## Industrial uses
Today most mercury is used in the primary separation of gold and silver ores prior to refining, in the manufacture of float plate glass, as a liquid metal electrode in the Solvay trough cell used for the manufacture of sodium hydroxide, and industrial catalysts. Some mercury is used in the construction of scientific laboratory instruments, switches, voltage rectifiers, X-ray tubes and photographic chemicals. Mercury is still used in the manufacture of small batteries.

## Detection methods
Colorimetric methods using di-beta-naphyl thiocarbazone in acid solution which gives a stable red complex have been successfully used for the estimation of total inorganic mercury.

Gas–liquid chromatography has been extensively used for the estimation of methyl mercury present in food. Homogenized food is treated with hydrochloric acid solution and the methylmercury chloride formed is extracted into benzene. An aqueous solution of cysteine is added to form a complex with the methyl mercury into the aqueous layer, which is then re-extracted into benzene. This extraction–re-extraction process is required to remove unwanted chemical compounds. The final benzene extract is analysed with a gas–liquid chromatography fitted with an electron capture detector.

Atomic absorption spectrophotometry is routinely used for the determination of trace quantities of mercury. Flame methods are suitable for concentrations as low as $0.2\,\mu g$ mercury/ml. Cold vapour flameless methods have been devised which convert the mercury present into the atomic form which can be liberated as a vapour. These methods are capable of detecting 10 ng mercury in a biological sample.

Mercury in air detectors are available commercially for the estimation of the concentration of mercury vapour present in the workplace. They range from:

Palladium chloride detector badges which change colour from yellow to grey in the presence of mercury vapour.

Chemical detector tubes based on the copper mercury complex formed when mercury reacts with copper iodide.

Portable atomic absorption monitors which can be used to determine the concentration of mercury vapour in the workplace by drawing air through a quartz tube which is placed in the path of an ultraviolet detector and measuring the mercury vapour absorption at 253.6 nm.

## Precautions when using the substance

For inorganic mercury compounds, particularly mercury $2^+$ compounds, acute poisoning can occur due to accidental ingestion of the salts or in the case of mercury metal by the inhalation of the vapour from the metal surface. Processes involving the hot working of ores or solutions are particularly hazardous. Mercury can produce allergic contact eczema and many of its salts are skin irritants.

Workplaces or laboratories where mercury or its compounds are handled must be subject to strict hygiene standards. Workers must wear personal protective clothing, including impervious gloves, and where possible all processes should be carried out in exhaust protective cabinets fitted with filters to prevent mercury dust or vapour being released to the environment. Where filtration cabinets cannot be used, personal respiratory protection masks fitted with filters designed for use with mercury, must be worn. Where mercury metal is used to fill instruments, thermometers or barometers, the work area must have provision for any mercury spillage to be contained in water containing spillage trays or troughs. In work areas having a high risk from mercury, regular monitoring of the air for the presence of mercury must be made and recorded. Levels found must be within the Occupational Exposure Standard.

## Occupational health

The main route of entry for mercury into the body is via the lungs. Mercury vapour present in inspired air is rapidly taken up from the alveoli into the circulating blood in the lungs, even at low mercury concentrations. Depending on the level of exposure, mercury in the blood exists as free mercury and some is oxidized to $Hg2^+$ which forms a complex with blood albumin and proteins in the red cells. The free mercury is a particular hazard as it is fat soluble and if present in significant amounts will cross the blood–brain barrier and deposit in the brain. Small amounts of free mercury may be oxidized to Hg2$^+$ and combine with proteins. The main sites of deposition of mercury metal following exposure are the brain and kidneys. The biological half-life of mercury in the body is 60 days and following a moderate exposure there will be an exponential release of mercury from the kidneys into the urine, and some excretion via the colon into the faeces. Urinary excretion of mercury shows a diurnal pattern being greater in the early morning than in the evening. For workers who regularly use mercury metal in their work, e.g. dentists and dental nurses, the urinary mercury level gives a good indication of occupational exposure. EOHS suggests that urinary levels <50 µg mercury/litre (compared with non-occupational exposure <10 µg mercury per day) indicate negligible risk from renal or neurological disorders.

Chronic poisoning usually starts slowly with clinical signs which are difficult to detect. Some neuromuscular and slight tremor may be seen with signs of proteinuria or albuminuria. If the exposure is not prevented, classical symptoms of heavy metal poisoning are seen with tremor, sweating, dermatography, diarrhoea and vomiting, leading to excessive loss of weight and significant proteinuria. Where the chronic poisoning is long standing, mercurial Parkinsonism is seen.

There are no specific antidotes for chronic poisoning. In cases of acute poisoning, various treatments have been suggested which increase the urinary excretion of mercury. These include calcium disodium versenate, penicillamine, $n$-acetyl D penicillamine, and British Anti-Lewisite or BAL (dimercaptopropanol).

## First aid

### Inhalation

If exposure to mercury vapour is suspected, the victim must be removed from the contaminated area into fresh air without delay, the rescuers using CABA equipment or gas masks designed for use with mercury. (Note: Gas masks should only be used in atmospheres containing normal oxygen concentrations, i.e. not basements or confined spaces).

Because persons affected by mercury vapour inhalation may subsequently develop lung inflammation it is important to ensure that the victim is placed in a raised position at rest and kept warm. If the casualty has difficulty breathing he/she should be given oxygen or oxygen-enriched air to breathe. The patient should then be taken to hospital for further treatment and the medical staff informed of the nature and if possible duration of exposure to the mercury vapour.

### Ingestion

Ingestion of any mercury compounds should be considered as serious, particularly if the compound is a mercury $2^+$ salt. If the patient is conscious, rinse the mouth out with cold water and give the victim half a litre of milk to drink to counteract the corrosive action of the ingested salt, and if possible encourage vomiting to remove the stomach contents. (If available give raw egg white to drink.) The victim should be kept warm and at rest, **and removed to hospital without delay for further treatment**.

### Skin contact

Mercury salts in contact with the skin should be rinsed away using large volumes of cold tap water, the washing process should be continued for at least 10 minutes. If no damage is caused to the skin surface, the patient should be seen by a medically qualified person to decide on possible further treatment. **If the skin is damaged or chemically burnt, the victim should be referred to hospital without delay for further treatment**.

### Eye contact

Wash away any mercury compounds from the surface of the eye by irrigating with a stream of cold tap water for at least 15 minutes, ensuring that the eye and orbit are fully cleaned. Place a dry eye pad dressing over the eye **and arrange for the casualty to be seen as soon as possible by an eye specialist at the nearest Accident and Emergency department.**

# Mercury organic compounds

Phenyl mercury acetate

$$C_6H_5HgCH_2COOH$$

Methyl mercury chloride

$$CH_3HgCl$$

Ethyl mercury chloride

$$C_2H_5HgCl$$

**Exposure limits**
Mercury alkyls (as mercury)
OES (TWA) = 0.01 mg/m$^3$
STEL = 0.03 mg/m$^3$

### Description
Phenyl mercury acetate is a crystalline material, slightly soluble in water but soluble in ethanol, benzene and acetone. Methyl and ethyl mercury chlorides are white crystalline substances soluble in ether.

### Properties
**Phenyl mercury acetate**

MW = 336.8, MP = 149°C.

**Methyl mercury chloride**

MW = 241, MP = 170°C.

**Ethyl mercury chloride**

MW = 265.1, MP = 192.5°C.

Because these compounds are easily taken into the body either as dusts, aerosols, or by ingestion, they must be considered as serious cumulative poisons with neurotoxic properties.

### Industrial uses
The main uses of organomercury compounds are as pesticides, fungicides and as a preservative for paints, manufactured products, and in the treatment of seeds. In the chemical industry, they are used as catalysts and the alkyls are used in some alkylating reactions.

## Detection methods

The best method for the identification and determination of aryl or alkyl mercury compounds is gas–liquid chromatography. Liquid samples should be pretreated with hydrochloric acid and extracted into benzene, followed by back extraction into aqueous cysteine solution to form a methyl mercury cysteine complex. Further acidification and re-extraction into benzene allows the alkyl mercury compound(s) to be separated and identified in the gas chromatograph using an electron capture detector.

Organomercury compounds present in the air can be estimated by passing a known volume of air through a gas scrubbing bottle, such as an Arnold bubbler, and analysing the contents of the liquid in the gas trap using the method described above.

## Precautions when using the substance

**Poisoning by alkyl mercury compounds can occur by inhalation of contaminated dust or vapour, or by skin contact with powdered compounds or concentrated solutions.** All workers handling these substances must wear protective clothing, neoprene gloves and respiratory protection if the process is not enclosed in an exhaust protective cabinet. Where workers are handling quantities of powdered material or spraying liquids, the operation must be carried out in an enclosed area with adequate exhaust ventilation protection. It is essential for workers to wear special chemical protective clothing and respiratory protection designed for use with mercury vapours and aerosols.

## Occupational health

Methyl and ethyl mercury compounds are more toxic to humans than phenyl mercury salts.

Data suggests that 95% of ingested methyl and ethyl mercury can be absorbed by the gastrointestinal system into the blood, compared with 2–10% for inorganic salts. Eighty per cent of volatile methyl mercury in inspired air is absorbed into the bloodstream; these fat soluble compounds are also easily absorbed through the skin, allowing alkyl compounds to readily enter the body. Methyl mercury is transported by the red cells in blood to the body tissues, especially in the central nervous system, where it concentrates in the grey matter particularly in the occipital cortex and the cerebellum. This may account for the severe ataxia and impairment of visual fields found in severe intoxications. Chronic exposure will also lead to sensory disturbances, parathesia in the extremities, lips and tongue, and auditory loss. These signs indicate neuronal damage to the visual and auditory cortex and the pre- and post-central areas of the brain verified by post mortem findings.

Onset of poisoning symptoms may be delayed depending on the extent of the exposure. The degree of exposure is best indicated by the methyl mercury or alkyl mercury concentration in blood. EOHS states that blood concentrations of 20–50 µg alkyl mercury/100 ml show slight intoxication whereas concentrations of 100–200 µg alkyl mercury/100 ml are associated with severe brain damage. The foetus is most at risk from exposure to organomercury compounds. It is two to five times more sensitive to the poison than adults. Because of this, great care and additional occupational health checks must be instituted when women are employed in factories handling or processing organomercury compounds.

Treatment of organomercury poisoning is more difficult than inorganic mercury intoxication because some of the chelating agents used, e.g. dimercaptopropanol, produce mercury complexes which can be transported across the blood–brain barrier and accumulate in the brain.

**First aid**

### Inhalation

The victim must be removed from the contaminated environment as soon as possible with the rescuers wearing respiratory equipment designed for use in mercury containing atmospheres. The victim should be placed at rest, kept warm, and given oxygen if breathing is affected. Do not use mouth-to-mouth resuscitation on the casualty, but, if necessary, use a face mask resuscitator fitted with an oxygen-enhanced air supply. **The patient should be removed to hospital without delay and seen by a medical specialist for assessment as to further treatment.** The medical staff should be informed of the type of organomercury compound involved and the duration of exposure suffered by the victim.

### Ingestion

**Ingestion of organomercury compounds is a serious health risk since 95% of the ingested material can be absorbed into the blood system unless effective treatment is given.** The patient may vomit and this can remove some of the toxic material from the stomach. If conscious, the victim should be given 500 ml of milk to buffer the poison and to delay transfer of stomach contents into the ileum. Keep the patient at rest and warm, **and transfer to hospital without delay informing hospital staff of the nature of the ingestion**.

Note for medical staff: Depending on the extent of the ingestion, a number of treatments are available to inactivate and remove the organomercury from the body. N-acetylpenicillamine and dimercaptosuccinic acid have been used to chelate with the organomercury and enhance urinary excretion of the complex. These chelating agents have been reported also to remove organomercury deposited in the brain. Because organomercury is excreted into the bile, stable ion exchange resins can be given by mouth which bind the mercury present in the gut and excrete it via the faeces.

### Skin contact

**Organomercury compounds can be rapidly absorbed if they come into contact with the skin or eyes.** Immediately irrigate the skin with a steady stream of cold or luke warm tap water. If the skin is unbroken wash with soap to remove as much of the spillage as possible, the first aider should wear rubber or vinyl gloves to prevent cross-contamination. The casualty should be referred to hospital for further medical assessment and treatment.

### Eye contact

Immediately irrigate the eye with a steady stream of cold tap water for at least 15 minutes ensuring that the cleansing process washes all surfaces of the eye and the surrounding tissues. Place a dry pad dressing over the eye(s) **and immediately refer the casualty to an eye specialist at the nearest Accident and Emergency department**.

# Metal fumes

**Exposure limits**
When the type of metal fume is known, the
OES of the metal vapour must not be exceeded
in the workplace.

## Description
Metal fumes arise from a number of metal production and fabrication operations.
They may be fumes or dusts arising from the tapping or pouring of molten metals
into moulds, or the heating, welding, or brazing of formed metal work giving rise
to metal fumes and toxic gases, such as oxides of nitrogen and carbon
monoxide.

## Properties
Fumes arising from the surfaces of molten or heated metals are of various particle
sizes. Those that enter the breathing zones of workers tend to be small, of the order
of 0.05–0.5 μm diameter, and they can act in the same way as gases entering the
respiratory system and penetrate into the lung alveoli.

## Precautions to be taken
**Many metallic particles and oxides can cause a medical condition known as
metal fume fever commonly occurring in industrial workers.** The toxicity of the
metal fumes depends on the metal used, e.g. cadmium fumes are very much more
toxic than iron oxide. The best precaution against the exposure to metal fumes is
to carry out pouring or cutting operations in well-ventilated areas or in the case
of welding work by supplying local exhaust ventilation which can remove dust,
fumes and toxic gases from the working surfaces. In some cases work has to be
carried out in specially ventilated enclosures, e.g. in the brazing of lead plates.
Where fumes are inevitable in the process, and ventilation cannot offer complete
protection, provision of respiratory protection either as full face gas or dust masks
will be necessary.

## Occupational health
Metal fume fever is a condition that shows itself in an acute and recurrent form.
Clinical symptoms, seen 4–8 hours following exposure, show as coughing,
dryness of the mucous membranes, weakness, fatigue, sometimes with severe
headaches and sometimes vomiting. Later on, 10–12 hours after exposure, the
patient exhibits fever with shivering, often with severe sweating, diarrhoea, and
frequency of urination. There is no specific treatment for metal fume fever except
where the metal concerned is known to be extremely toxic, e.g. mercury, where
specific treatments are available and are described in the relevant sections in this
book.

**First aid**

No specific treatment is available for metal fume fever. The patient should be sent home or removed from the exposure, kept warm, and treated as for influenza with warm drinks and rest. Complete recovery is usually seen within 24 hours following exposure.

# Molybdenum and compounds

# Mo

### Exposure limits

| | |
|---|---|
| Insoluble compounds (as Mo) | OES (TWA) = 10 mg/m$^3$ |
| | STEL = 20 mg/m$^3$ |
| Soluble compounds (as Mo) | OES (TWA) = 5 mg/m$^3$ |
| | STEL = 10 mg/m$^3$ |

## Description

Molybdenum exists as a silver-white metal or as a dark grey-black powder. Molybdenum exists as a large range of commercially useful compounds because the element exists in six valency states.

## Properties

Molybdenum metal is soluble in aqua regia, and hot concentrated nitric or sulphuric acid.

The element exists in six valency states ranging from the free metal Mo(0) to Mo(6$^+$). This feature enables molybdenum to form many compounds and complexes which are useful commercially.

## Industrial uses

Molybdenum compounds are widely used in the manufacture of catalysts in specialized steel manufacture. Molybdenum sulphide is used as a heat resistant lubricant additive as the sulphide readily deposits on hot metal surfaces to form a heat resistant lubricant ('molyslip'). Molybdenum hexacarbonyl is used to produce organomolybdenum dyes and molybdenum thermal plating. Some soluble molybdenum salts are used as a trace metal supplement in agriculture.

## Detection methods

Colorimetric analysis using reagents such as thiocyanate, phenylhydrazine or chloranilic acid which form coloured complexes.

DC arc spectrographic analysis has been used to determine molybdenum in foodstuffs, and more recently in serum, urine and hair.

Atomic absorption spectroscopy using an air/acetylene flame following wet oxidation of the sample and chelation of the molybdenum using ammonium pyrollidine dithiocarbamate, and extraction of the complex into methyl-isobutylketone.

More recently, carbon furnace atomic absorption spectroscopy using the chelation/ extraction technique described has produced a greater analytical sensitivity.

## Precautions when using the substance

There are considerable differences in the toxicity of molybdenum compounds. Insoluble molybdenum salts, e.g. molybdenum sulphide and the halides have low toxicity, but the anionic soluble compounds, e.g. molybdates are quite toxic and must be handled with care.

**In industrial situations where insoluble molybdenum compounds are being used, the greatest dangers will occur if the workers are exposed to high levels of dust, fumes or mists containing molybdenum. Dust and fume suppression at the workplace is important.**

## Occupational health

Although molybdenum compounds are used a great deal in industry, there are few serious toxic effects reported from their use.

Molybdenum trioxide, which volatilizes above 800°C, can produce acute exposure symptoms of irritation to the eyes, throat and mucous membranes. Molybdenum toxicity, not infrequently found in cattle grazing on molybdenum rich pastures, will in time produce anaemia, and a form of gastrointestinal diarrhoea known as 'scouring'. This can be treated effectively with oral copper sulphate solutions, since copper and sulphate ions counteract the harmful effects of molybdenum.

EOHS reports an interesting correlation between the handling of soluble molybdenum solutions by chemists and a high frequency of gout. Investigations show that there is strong correlation between the molybdenum content of food, the incidence of gout, uric acid concentration in the blood, and its xanthine oxidase activity.

---

### First aid

#### Inhalation
Persons who are working in industrial situations and who have suffered acute exposure to molybdenum containing fumes or dust, should be removed from the contaminated area into fresh air and a first aid assessment of the casualty made. If the pulse and breathing are normal, the patient should be rested and observed to ensure no further symptoms develop. If in doubt about the condition of the patient, he/she should be examined by a medical doctor.

#### Ingestion
Ingestion of soluble molybdenum salts usually causes nausea and mild abdominal pain. Provided the patient is conscious and not likely to collapse, let the patient vomit up any toxic material and then encouraged them to drink a pint of milk or water. Milk will buffer any acidic material and also delay the stomach emptying time into the duodenum. The patient should be taken to hospital for a gastric lavage to wash out the remaining toxic salts.

#### Eye contact
Solid material or solutions spashed into the eyes should be treated by immediately irrigating the eyes with a slow stream of cold tap water for at least 15 minutes, preferably using a flexible hose to direct the water into the orbit of the eye. The patient should be encouraged to open and shut the eyes during this process. As soon as the irrigation process has been completed, the eye(s) should be covered with a dry loose dressing **and the casualty taken to hospital without delay to receive specialist treatment.**

---

# Nickel carbonyl $Ni(CO)_4$

**Exposure limits**
OES = 0.001 g/m³ (0.007 mg/m³)
STEL = no values quoted

## Description

A colourless, highly flammable liquid which produces a very heavy vapour with a musty odour. Usually supplied as a liquid in cylinders pressurized with carbon monoxide.

## Properties

A mobile colourless liquid, soluble in organic solvents which decomposes at 60°C producing carbon monoxide and finely divided nickel. It slowly oxidizes in air.

MW = 170.75,  SG = 1.32,  MP = −25°C,  BP = 43°C.

## Detection methods

Infra-red gas analyser.
Methylated spirit flame test. The presence of nickel carbonyl is seen as a yellow cap on the surface of the blue methylated spirit flame. This test is sensitive to one part of $Ni(CO)_4$ in 400 000 of air.

## Precautions when using the gas

Nickel carbonyl is one of the most dangerous substances used in industry in large quantities.

It is extremely hazardous because the acute toxic level for humans is low. It is a human carcinogen, extremely flammable, and because the vapour is six times heavier than air dispersal of the gas, if released, is difficult.

Braker *et al.* (1977) give good advice on the precautions to be observed in industry:

1. Nickel carbonyl systems must be operated completely closed to the atmosphere.

2. Pipework in the plant using the gas must be positive pressure tested to twice the working pressure with nitrogen before use.

3. No nickel carbonyl gas should be burnt off or discharged to atmosphere.

4. Full safety precautions and breathing equipment suitable for handling a lethal gas must be available at all times.

5. Accident and emergency schemes together with a full risk assessment must be prepared and tested prior to the operation of any plant.

## Occupational health

The formula of nickel carbonyl shows that it has two toxic components, metallic nickel and carbon monoxide. The effect of this compound on man is reflected on the toxicity of both these components. Nickel carbonyl normally enters the body through the lungs, but absorption through the skin is possible.

Acute effects of exposure can be considered as initial symptoms due to the inhalation of carbon monoxide and later or delayed symptoms due to the toxic effect of absorbed nickel.

Initial symptoms of poisoning are headache, dizziness, nausea and vomiting followed by tightness of the chest. Later, sweating and fever may occur.

Delayed symptoms occurring 12–36 hours after the initial exposure involve metallic taste and severe retrosternal pain. Any damage to the pulmonary capillary epithelium will cause oedema and capillary haemorrhage and result in a cough with haemoptysis.

**If untreated, severe poisoning can lead to degenerative changes in the liver, kidneys and central nervous system.**

Severity of exposure can be assessed by the pulse:respiration ratio. Any case with a ratio of less than 3:1 (normal 4:1) should be carefully monitored. A knowledge of the blood and urine nickel concentrations is valuable.

Long-term effects of exposure to nickel carbonyl are eczematoid dermatitis, carcinoma of the nasal sinuses and lungs, and chronic asthma.

**First aid**

EXPOSURE TO NICKEL CARBONYL LIQUID OR GAS IS A MEDICAL EMERGENCY.

It is vital to carry out the following procedures without delay.

1. Using positive pressure breathing apparatus and protective clothing, remove victim from the toxic atmosphere to fresh air.

2. Remove all of the casualty's outer clothing and place into a thick plastic bag and seal. (This must be done while the rescuer is still wearing chemical protective clothing and breathing apparatus to protect the rescuer from any vapour or liquid on the victim.)

3. If the subject is breathing, administer a gas mixture of 95% oxygen/5% carbon dioxide. If the subject is not breathing, administer this gas mixture with artificial respiration preferably using an Ambubag. NO MOUTH-TO-MOUTH RESUSCITATION. Keep the subject warm and at rest preferably on a stretcher. A urine sample should be collected if possible and kept with the patient.

4. Call a medically qualified person as soon as possible giving details of the type of poisoning suspected.

THIS IS A MEDICAL EMERGENCY – THE PATIENT SHOULD BE TAKEN TO HOSPITAL WITHOUT DELAY. A label giving the time and type of treatment given should be attached to the wrist.

## Notes for medical staff dealing with nickel carbamyl poisoning

The classification of the degree of exposure is best made with reference to the severity of clinical symptoms already described and the nickel concentration found in the first 8-hour urine sample.

The type and extent of the treatment needed is related to this urine level. One of the most effective chelating agents for nickel is sodium dithiocarbamate (dithiocarb).

### Summary of clinical chemistry related to exposure

Initial 8-hour urine sample less than $10\,\mu g$ nickel/100 ml and/or a blood nickel concentration of 3–10 mg per litre.

Exposure is slight. Delayed symptoms will be absent or slight. The patient should have a daily urinary nickel measurement made until the level has returned to the normal range of 0–3 $\mu g$/100 ml.

If the urinary nickel level is >10 but <50 $\mu g$/100 ml exposure has been severe, the patient will be seriously ill and require hospitalization and chelation therapy. Treatment should consist of oral sodium dithiocarbamate (0.5 grams of dithiocarb with 0.5 grams of sodium bicarbonate) four times a day with 500 ml of water. Treatment should be continued until the patient is free of symptoms and the urinary nickel concentration is <10 $\mu g$/100 ml.

If the patient's condition appears critical, dithiocarb may be given parenterally. An initial dose of 25 mg dithiocarb/kg body weight up to a maximum of 100 mg dithiocarb/kg body weight/day. Dithiocarb is prepared for injection by adding 10 ml of sterile phosphate buffer (0.5 grams of monosodium phosphate in 100 ml of sterile water) to 1 gram of dithiocarb – this solution contains 100 mg dithiocarb per ml. Sixteen days of treatment may be required before the nickel concentration in the patient's urine is normal.

# Nitrogen dioxide NO$_2$
# Dinitrogen tetroxide N$_2$O$_4$

**Exposure limits**
OES = 3 ppm (5 mg/m$^3$)
STEL = 5 ppm (9 mg/m$^3$)

## Description
**A reddish brown gas with an irritating and poisonous vapour.** It is produced from nitric oxide when present with air or oxygen and reactions involving concentrated nitric acid.

## Properties
The gas is soluble in carbon disulphide, alkalis and chloroform. It is one component of the so-called 'nitrous fumes' evolved during acid pickling or the oxidation of organic matter with nitric acid.

MW = 46, SG = 1.45, MP = −11.2°C, BP = 21.2°C,
VD = 1.58, VP = 760 mmHg (101.10 × 10$^3$ Pa)

## Detection methods
Chemical reaction tubes.

## Precautions when using the gas
Nitrogen dioxide and dinitrogen tetroxide exist in a temperature dependent dynamic equilibrium and the precautions for these two forms is similar. The gases are easily produced in chemical reactions involving nitrates and mineral acids, reactions with nitric acid, and as a result of the operation of internal combustion engines. **Because of the low OES ascribed to nitrogen dioxide, work with processes that can generate this gas must be carried out in well-ventilated areas**, e.g. under fume extraction hoods or, in the case of chemical laboratories, in an efficient exhaust protective cabinet. Remember that both gas and arc welding can produce quantities of nitric oxide and nitrogen dioxide. **Staff who regularly work in atmospheres that contain nitrogen oxides should wear suitable chemical resistant clothing, face and hand protection. In addition, personal sampler or indicator tubes should be worn, so that regular checks can be made of the level of exposure to this toxic gas.**

In workplaces where nitrogen dioxide is used, eye wash stations and showers should be available and self-contained breathing apparatus for rescue by trained staff in case of an emergency.

## Occupational health

Nitrogen dioxide is a powerful lung irritant and is thought by most experts to be the oxide of nitrogen responsible for the hazards from 'nitrous fumes' exposure. The EOHS suggest that **exposure to gas concentrations of 100–500 ppm may lead to sudden death from bronchospasm and respiratory failure**. **Death can also result from delayed respiratory failure or inflammatory changes caused by a form of bronchiolitis possibly caused by an autoimmune response to the exposure.** There are also possibilities that moderate exposures of the order of 20–50 ppm can give rise to chronic respiratory disease. Because of this data any exposure to nitrogen dioxide unless of a trivial nature must be taken seriously by management.

---

**First aid**

Inhalation
**Prompt treatment of a person exposed to nitrogen dioxide vapour is essential to prevent delayed, serious and sometimes fatal damage to the lungs. Remove the victim from the contaminated area (the rescuers wearing full face compressed breathing apparatus) into fresh air.**

If conscious, instruct the casualty to breathe deeply to remove as much nitrogen dioxide gas from the lungs as possible. Then administer oxygen-enriched air from an oxygen-supplemented resuscitator, not **mouth-to-mouth resuscitation** until the casualty's skin colour shows normal oxygenation of the tissues.

If unconscious, **this is a medical emergency**, ventilate the patient, lying down and at rest, with an oxygen-supplemented resuscitator, **do not use mouth to mouth method**, until the patient revives. Keep the casualty at rest and warm and supply further oxygen-enriched air to maintain a normal oxygenated skin colour. Do not leave the patient. Call for assistance from a medically qualified person and then arrange for the patient to be removed to hospital.

**Note: The supply of oxygen at the early stages of the treatment for these patients is vital to reduce the development of pulmonary oedema and/or inflammation of the airways produced by this corrosive gas.**

Skin contact
Place the victim under an emergency water shower or drench with a hosepipe removing all contaminated clothing. Wash the affected skin area(s) with a flow of cold tap water, using soap if the skin is unbroken, for at least 15 minutes. Cover any burns to the skin with a loose dry dressing and remove the casualty to hospital.

Eye contact
**Severe corrosive damage can occur if the eyes are in contact with nitrogen dioxide gas or liquid.** The patient should be removed from the contaminated area as described above and the eyes thoroughly irrigated with cold tap water for at least 15 minutes making sure that the surfaces of the eyelids are properly cleaned. Patients will object to this procedure and will tend to keep their eyelids closed particularly if some liquid has entered the eye. Insisting on thorough treatment could save the patient's sight. Repeat the washing process if severe pain is still felt by the casualty. **Arrange for the patient to be removed to hospital as soon as possible to be treated by an ophthalmic expert.**

---

# Nitrogen trifluoride     NF$_3$

**Exposure limits**
OES = 10 ppm (30 mg/m$^3$)
STEL = 15 ppm (45 mg/m$^3$)

## Description
Nitrogen fluoride is a colourless, toxic gas with a characteristic mouldy smell. It is usually supplied in small cylinders at 200 psi.

## Properties
**This reactive toxic gas can be detonated by exposure to heat, flame, electric sparks or organic material and supports combustion.** It is slightly soluble in water with reaction.

MW = 71,   SG = 1.537 (liquid),   MP = −128.8°C.

## Detection methods
Infra-red gas analyser.

## Precautions when using the gas
**Because of its reactive properties in the presence of oil or grease, all equipment in contact with nitrogen trifluoride must be carefully cleaned.** The gas cylinders should be stored in a well-ventilated and secure place and provided with easily available dry powder fire extinguishers. The gas cylinders should only be used in mechanically well-ventilated areas or in exhaust protective cabinets. Chemical resistant clothing, with eye, hand and face protection, similar to the type used for handling hydrogen fluoride, must be worn.

## Occupational health
As nitrogen trifluoride is a toxic gas used mainly in synthetic chemical reactions, **it should only be used by staff who have been trained in the handling and operation of the equipment, and in the safety procedures required in the event of a serious leakage of gas. Preplanning and a full risk assessment must be made before any work is attempted.** Nitrogen trifluoride present in respirable air at concentrations above the OES of 10 ppm can cause lung irritation, coughing, headache, nausea and diarrhoea.

**First aid**

Inhalation

The subject should be removed as soon as possible into fresh air and 100% oxygen administered by means of an oxygen resuscitator. Even with minor exposures, 100% oxygen should be given at intervals up to 4 hours following exposure. Medical advice should be sought as to any further treatment required.

**In severe exposures, the victim must receive oxygen therapy immediately, and then removed by ambulance to hospital without delay.** Seek advice from a medically qualified person if available. A member of staff should accompany the casualty to hospital to give details to medical staff of the type and duration of exposure to assist in further treatment.

Skin contact or clothing saturated with the gas

Immediately drench the affected area or whole body with a shower of cold water. Remove the victim's clothing while under the shower. Continue cold water treatment for at least 15 minutes to remove as much fluoride from the surface of the skin as possible. Get an assistant to seek medically qualified assistance as soon as possible. If this is not possible arrange for the patient to be removed to hospital by ambulance. Continue to treat the subject with first aid which will reduce the absorption of the fluoride through the skin.

Treat the affected area with 70% iced ethyl alcohol (ethanol) or an iced solution of magnesium sulphate for 30 minutes. If medical aid is delayed, continue the iced ethanol or magnesium sulphate until help arrives. If medical aid cannot be obtained, apply a 10% calcium gluconate gel to the affected area of skin.

Eye contact

THIS IS A MEDICAL EMERGENCY.

**Immediately irrigate the eyes with a steady stream of cold tap water for at least 15 minutes making sure that the eye and surrounding tissues are well washed.** Place a loose dry dressing over the eyes **and immediately remove the casualty to hospital for further specialized eye treatment, informing the medical staff of the name and nature of the chemical involved.**

# Nitrosyl chloride    NOCl

## Exposure limits
No OES have been quoted, presumably because
of its toxicity.

## Description
An orange-yellow highly toxic gas with an irritating penetrating smell.

## Properties
The gas is non-flammable, decomposing easily in the presence of water or moisture to give hydrochloric and nitrous acids. It is easily liquefied and supplied as the liquefied gas in steel cylinders.

MW = 65.6,   BP = –5.8°C,   SG = 1.4 at –12°C,   MP = –64.5°C.

## Detection methods
White fumes in the presence of filter paper moistened with ammonium hydroxide solution.
Chemical detector tubes designated for nitrous fumes.

## Precautions when using the gas
**This is a dangerous substance and must be treated with care. All persons intending to use this gas must carry out a detailed risk assessment and ensure that the appropriate safety measures and equipment are available and in working order.**

Purchase and store only essential amounts of nitrosyl chloride. The gas cylinders must be stored in a well-ventilated secure area protected from the weather.

**Any work or process involving nitrosyl chloride gas must be carried out in a well-ventilated hood or exhaust protective cabinet.** Because of the corrosive nature of the gas and its hydrolysis to hydrochloric and nitrous acids, pipework and vessels used must be made of monel, nickel metal, or Teflon, and carefully cleaned and dried before use. **An antisuck-back trap must be fitted to the cylinder valve to prevent water or moisture entering the cylinder and causing a dangerous build-up of pressure.**

**Persons using equipment containing nitrosyl chloride must wear full protective chemical suits with chemical resistant gloves and boots, and full eye protection, e.g. full face shield with polycarbonate goggles. An eye wash station and shower must be accessible.**

## Occupational health
The toxicity of this gas has not be fully established in humans. Animal tests show that concentrations of the order of 100 ppm for 10 minutes can be lethal, due to fatal

pulmonary oedema. The gas is known to be highly irritant to the skin, eyes, and mucous membranes at much lower concentrations. It is suggested that if the gas can be detected by smell, this will indicate a dangerous level of the gas.

---

### First aid

#### Inhalation
**Remove the casualty form the affected area into fresh air at once, the rescuers wearing compressed air breathing apparatus, not chemical filtration respirator.** Lay the patient down and keep him/her warm and at rest. If the patient is not breathing, administer 100% oxygen by means of an oxygen-supplemented respirator. If the patient is conscious, try to discourage the patient from coughing, and then use an oxygen-enriched artificial ventilator until breathing is normal.

#### Skin contact and contaminated clothing
Provided the patient is breathing normally, drench the affected skin area or clothing with a stream of cold water, removing the clothing during this process. Continue the irrigation process for at least 15 minutes. If the patient stops breathing, remove from the shower and start artificial respiration with supplementary oxygen.

#### Eye contact
**THIS IS A MEDICAL EMERGENCY.**

Immediately treat the casualty by irrigating the eye with a stream of cold tap water ensuring that the whole of the eye and surrounding tissues are fully washed. Maintain this treatment for at least 15 minutes. Repeat the treatment if you suspect that there is chemical still remaining in the eye. Cover the eye with a dry eye pad dressing. **Do not put any medication into the eye and then take the patient to the nearest casualty hospital without delay.** Inform the medical staff at the hospital of the nature of the exposure and the substances involved.

---

# Nitrous oxide

# N₂O

## Exposure limits
No OES quoted for this gas. Full anaesthesia is only obtained when the concentration of nitrous oxide in air breathed exceeds 70% by volume. Dentists use lower concentrations with oxygen to produce relative analgesia for nervous patients.

## Laughing gas

## Description
A colourless gas possessing a pleasant sweetish odour. Used with oxygen as an anaesthetic gas in operating theatres.

## Properties
The gas is soluble in water, ethyl alcohol, and diethyl ether. Because of its high water solubility and low toxicity, it is used in the food manufacturing industry as a propellant for pressurized cans.

$MW = 44$,  $SG = 1.98$,  $MP = -90.8°C$,  $BP = -88.5°C$,  $VD = 1.53$,
$VP = 760 \, mmHg$ ($101.10 \times 10^3$ Pa) at 88.5°C.

## Detection methods
Infra-red analysis.

## Precautions when using the gas
The anaesthetic property of the gas means that care must be taken to ensure that any work with nitrous oxide is carried out in a well-ventilated workplace. Very high concentrations of the gas are not in themselves toxic but tend to reduce the available oxygen content of the breathable air. Research suggests that dentists and anaesthetists who are exposed to nitrous oxide in the frequent testing of anaesthetic masks, or addictive gas sniffing, may be at risk because the inhaled nitrous oxide can oxidize vitamin $B_{12}$ causing pernicious anaemia and subacute combined degeneration of the spinal cord. This problem can be easily rectified by giving small amounts of vitamin $B_{12}$ supplements in the diet. The requirement of this vitamin for humans is only 1 microgram per day.

## Occupational health
No special health considerations are necessary as the gas is regarded as non-toxic.

---

**First aid**

In cases of anaesthesia due to inhalation of pure nitrous oxide gas, the subject should be removed from the contaminated area to fresh air and pure oxygen administered until breathing is normal. Keep the patient at rest and warm until fully recovered.

---

# Osmium and compounds

# Os

**Exposure limits**
Osmium tetroxide (as Os)
OES (TWA) = 0.0002 ppm (0.002 mg/m$^3$)
STEL = 0.0006 ppm (0.006 mg/m$^3$)

## Description
Osmium is a greyish blue, lustrous and brittle metal, produced by a naturally occurring alloy osmiridium in the mining of platinum.

## Properties
Osmium readily forms alloys with the platinic metals and iron cobalt, nickel, tin and zinc. The metal easily oxidizes in air to form osmium tetroxide which is very volatile and poisonous. Osmium tetroxide is a volatile and strong oxidizing agent forming osmium dioxide and finally metallic osmium in its reactions. These properties of oxidation and staining are used extensively in the preparation of tissue specimens for electron microscopy. Osmium also readily reacts with chlorine and carbon monoxide to form chlorides and carbonyls respectively.

## Industrial uses
Osmium is used as an industrial catalyst in the synthesis of ammonia and organic compounds. The omiridium alloy is used in the coating process for fine bearings, ink pens, and compass needles.

## Detection methods
Colorimetric method using thiourea which forms a coloured product with osmium.

The presence of osmium tetroxide vapour can be indicated by oxidation of a neutral fat stain on exposed filter paper forming black hydrated osmium dioxide and osmium metal.

## Precautions when using the substance
As indicated by the Occupational Exposure Standards, osmium tetroxide and osmium containing vapours are extremely poisonous and harmful to the eyes. Osmium is one of the most toxic materials used in industry and a full risk assessment must be made prior to any work involving osmium or its compounds.

Fortunately, osmium and osmium tetroxide have a very irritating smell reminiscent of bromine vapour. The vapour from a 1% solution of osmium tetroxide can damage the cornea of the eyes by its powerful oxidative action and this results in a deposition of osmium dioxide and osmium metal producing a black deposit over the eye leading to blindness.

Very low concentrations of the osmium containing vapour will cause extreme irritation of the eyes, conjunctivitis, and damage to the upper respiratory system, leading to breathing difficulties, bronchial spasms and bronchitis.

**All work using osmium must be carried out in a designated area with full local exhaust ventilation where the extracted air is filtered to remove any osmium. Workers using osmium must wear protective clothing, hand protection, and full face respiratory protective equipment. If possible, the work should be confined to a gas tight protective cabinet fitted with independent filtration facilities.**

The smell associated with osmium tetroxide should be a warning of the presence of a toxic level in the air and all staff should be evacuated immediately and not return until the area has been decontaminated and declared safe.

**First aid**

Any person affected by osmium vapour must be treated as a medical emergency. The affected persons must be removed from the contaminated area into fresh air, the rescuers using compressed air breathing apparatus and chemical protective clothing including gloves.

### Inhalation

Mild inhalation of osmium causes a headache, soreness of the throat, and a tight feeling in the chest. Serious inhalation of vapour will produce severe breathing difficulties.

The victim must be placed at rest, kept warm to reduce any shock and given positive pressure ventilation with an oxygen-enriched resuscitation mask to reduce the possibility of pulmonary oedema. The casualty should be removed to hospital as soon as possible, with a member of staff in attendance to inform the Accident and Emergency department of the hospital of the nature and possible extent of the exposure.

### Ingestion

Any ingestion of osmium vapour is likely to produce nausea and vomiting. If the victim is conscious, get them to drink a glass of milk (or water). Milk will reduce the toxicity of the osmium salts and delay the emptying of the stomach contents into the duodenum. This will allow the remaining osmium to be removed by gastric lavage at the nearest hospital. **As soon as possible, arrange for the patient to be taken to hospital for further specialist treatment.** Failure to do this may result in serious kidney damage to the victim.

### Skin contact

Any osmium compounds splashed onto the skin must be quickly removed with a copious stream of cold tap water, irrigating the affected area for at least 10 minutes. The skin may appear very red and irritated at the contaminated area. Following the irrigation treatment, the skin should be covered with a clean loose dressing, and the patient referred to hospital for specialist treatment.

### Eye contact

THIS IS A MEDICAL EMERGENCY.

As soon as possible, irrigate the eye(s) with a slow stream of cold tap water making sure that the whole eye and orbit are well washed to remove any remaining chemicals. Place a dry loose dressing over the eye(s) **and immediately refer the casualty to hospital for specialist treatment. Irreversible corneal damage can occur if the deposition of osmium onto the cornea is not prevented.**

# Oxygen  O$_2$

## Description

A colourless, odourless and tasteless gas, also existing in the form of a light blue liquid.

## Properties

Oxygen as a gas in the air is essential to the life of aerobic organisms. It can exist as reactive atomic oxygen, the stable diatomic oxygen, and the reactive ozone. Ozone because of its special hazards will be considered separately. Oxygen reacts with most elements to form oxides. The diatomic gas is an essential component in the production of fire.

MW = 32,   SG = 1.14 (liquid),   MP = –218.4°C,   BP = –183°C,   VD = 1.43.

## Detection methods

Paramagnetic balance method.
Polarographic cell method.
Chemical gas detector tubes.

## Precautions when using the gas

Oxygen gas is supplied in cylinders of various sizes and capacities, and is used for medical respiratory purposes to oxygen cutting equipment in industry. The pressures of these cylinders can be from 150 to 160 atmospheres. Pure oxygen under pressure escaping from a cylinder in contact with organic materials can lead to spontaneous heating and ignition. Liquid oxygen, in addition to the dangers associated with a cryogenic fluid, can explode if placed in contact with oxidizable materials. Handling of liquid oxygen requires protective clothing, special gloves and face and eye protection, particularly when pouring the liquid into Dewer type receiver vessels. Liquid oxygen tanks are now found both in industry and as a source of medical oxygen in large hospitals, where supplies of oxygen are drawn from the evaporation of the liquid into supply pipelines to the wards.

Oxygen cylinders and liquid oxygen tanks must be sited so that they are in a well-ventilated area. In the case of liquid oxygen the tank must be provided with bunding sufficient to contain the liquid if, in the event of an accident, the tank ruptured. Where oxygen pipelines are installed, automatic shutdown facilities must be provided in case of fire.

## Occupational health

Both an excess and deficiency of oxygen is harmful to the body. Oxygen deficiency in the air is often found in mines, sewers and confined spaces. Work should not be carried out in atmospheres where the oxygen concentration is less than 18% by volume. Workers in confined spaces should be provided with portable low oxygen alarms to warn of danger. No person should be permitted to work alone in these conditions.

Inhalation of pure oxygen uder pressure is harmful and causes pulmonary inflammation and haemorrhage leading to fatal pulmonary oedema.

---

**First aid**

Inhalation
Workers in oxygen-rich atmospheres should be removed to fresh air and allowed to recover and the rescuer should remain with the patient until the respiration depth and rate is normal. Low carbon dioxide level in the lungs can produce abnormal respiration rates with periods of cessation of breathing (apnoea). Persons who have been rescued from an oxygen deficient atmosphere should be given 100% oxygen from a medical cylinder or oxygen-enriched air from a resuscitator until the breathing is normal.

Skin and eye contact with liquid oxygen
THESE ARE MEDICAL EMERGENCIES.

Immediately irrigate the affected area with cold water for at least 15 minutes. Instruct an assistant to call for medically qualified assistance and an ambulance as soon as possible. The casualty will require expert accident and emergency treatment in the nearest hospital.

---

# Ozone $O_3$

## Exposure limits
OES = 0.1 ppm (0.2 mg/m$^3$)
STEL = 0.3 ppm (0.6 mg/m$^3$)

## Description
A bluish gas with a characteristic smell associated with electrical discharges or arc welding. **It is an extremely powerful oxidizing agent.**

## Properties
Ozone is an extremely reactive gas which is slightly soluble in water and very soluble in alkalis and oils. Its strong oxidizing powers are used in the textile industry for the bleaching of fabrics and in the deodorizing and purification of water and air.

MW = 48,  SG = 2.1 (gas) 1.6 (liquid),  MP = −192.7°C,  BP = −111.9°C,  VD = 1.66.

## Detection methods
Chemical gas detector tubes.
Ozone electronic gas detectors.
Potassium iodide impregnated papers.
Release of titratable iodine from a potassium iodide solution in buffered phosphate.

## Precautions when using the gas
**The very low OES quoted for ozone indicates the high toxicity of the gas in the working environment.** Ozone can be easily produced as a result of electrical discharges or from electrical inert gas shielded arc welding where ozone concentrations of up to 9 ppm have been reported. It is very important that all operations where there is a possibility of ozone production must be carried out in well-ventilated areas. (Note: Some photocopying machines produce ozone during their use and they should not be sited in unventilated anterooms or cupboards.)

**Ozone will react violently with many metal catalysts, in some cases producing explosions. Liquid ozone is particularly unstable and explosive. Ozone will also react with unsaturated organic compounds to produce very unstable and explosive ozonides.**

Cylinders containing dissolved ozone must be stored separately in a cool or refrigerated area well away from reducing agents or ozone catalysts.

Work with ozone in manufacturing industry requires special exhaust protective ventilation systems which incorporate air scrubbers that remove ozone from extracted air.

Work with liquid ozone must only be undertaken by staff specially trained in its handling and use. In addition to the hazards mentioned, liquid ozone is a cryogenic liquid and all the precautions associated with extreme cold exposures must be taken.

## Occupational health

Ozone is a highly toxic irritant gas and in view of the very low Occupational Exposure Standard quoted for this gas, workers must not be exposed to it. Accidental exposure from a short time causes oxidative damage to the respiratory system indicated by inflammation and congestion of the lungs. In more severe exposures, acute pulmonary oedema, haemorrhage and death have been reported. Fortunately the level of detection of the gas by smell is 0.05 ppm in air so that slight leakages can be detected easily and the area evacuated until proper ventilation and repairs to leaking equipment have been made.

Recently an interesting similarity has been suggested between the toxic effects of ozone and the health hazards associated with ionizing radiations. The link derives from the fact that ozone behaves as a free radical, and a number of theories associate biological damage caused by ionizing radiation to the production of free radicals in body tissues. The EOHS state that toxicologically, ozone and ionizing radiation have at least six actions in common. This radiomimetic nature of ozone means that no level of exposure to the gas is safe and that the OES exposure level should not be regarded as a safe working concentration.

---

**First aid**

ANY EXPOSURE TO OZONE GAS OR LIQUID MUST BE CONSIDERED AS A MEDICAL EMERGENCY.

Inhalation

**Remove the victim from the contaminated area into fresh air without delay; the rescuers must wear compressed air breathing apparatus.** If the casualty is breathing give oxygen-enriched air (or 100% oxygen) from a resuscitator until breathing is normal. If the victim is unconscious, immediately use an oxygen-enriched resuscitator to revive the patient. Ask an assistant to call for an ambulance, do not leave the casualty, and if possible obtain medically qualified advice. Continue to treat the patient for possible congestion of the lungs. Inform the ambulance staff of the nature and, if possible, duration of the exposure.

Skin and eye contact

If ozone liquid or gas contact is suspected, irrigate the affected parts with large volumes of cold tap water for at least 15 minutes, removing any clothing which may contain the chemical. While this treatment is in progress, arrange for an assistant to call for an ambulance to remove the casualty to hospital for specialist treatment. Liquid ozone is extremely toxic and is a cryogenic fluid causing serious cold burns and tissue damage.

---

# Petroleum and products

## Petroleum spirit, motor spirit, gasoline

**Exposure limits**
No OES or STEL quoted by the HSE

### Description
Petrol or petroleum spirit is a highly flammable volatile colourless organic liquid. Petrol vapour can form a very explosive mixture with air.

### Properties
Petrol is a liquid containing long chain paraffins and aromatic hydrocarbons used as a fuel for internal combustion engines. For use in high compression engines, additives are blended with petrol to give a fuel which reduces pre-ignition of the fuel in the engine cylinders. The most common additive was tetraethyl lead, which is still used in 'leaded petrol'. With the advent of lead-free environmentally cleaner fuels, this additive has been replaced by aromatic compounds such as triorthocresyl phosphate, known in the USA as TCP. Octane ratings of petrol can be raised by adding a mixture of benzene, toluene and xylene, known as benzol, to the basic fuel.

Petroleum spirit can be supplied in distilled fractions with a BP in the range 30–160°C.

FP = –17°C for the lower fraction, EL = 1–6%, IT = 290°C.

### Industrial uses
The main use of petrol is as a fuel in the motor transport and aircraft industries. Higher fraction petroleum products are used as paraffin (kerosene) for cooking and lighting and as aviation jet engine fuel. Lower fraction petroleum is used as a solvent in the rubber, paint and adhesive industries.

### Detection methods
The presence and concentration of petrol vapour are measured by means of a flammable gas analyser using a fuel cell detector. **Any instrument of this type used for the detection of petrol vapours must be designed to work in explosive atmospheres.** Some manufacturers produce portable multiwarning systems which are able to measure flammable, toxic gases and oxygen concentration.

### Precautions when using the substance
**The greatest dangers from petrol are explosion and fire.** Petrol vapour can easily be formed in enclosed spaces or containers even at low ambient temperatures. Ignition can be caused by electrical sparks or static discharges as well as naked flames.

Special care must be taken in the transport and storage of petroleum spirit in industry where large volumes are regularly handled.

**Smoking and the use of any naked flame must be prohibited in the work or storage area where petroleum products are held or used.**

Storage tanks or vessels that need cleaning out to remove sludge must be completely emptied out and ventilated or purged with inert gas before cleaning. Catastrophic explosions have occurred in oil tankers at sea when tanks have been cleaned out using water hoses. The water jets have produced a build-up of static electricity inside the tank, leading to a static spark discharge and explosion.

Small or large storage tanks that need welding repairs must be carefully emptied and petrol vapour purged from the tank using an inert gas, e.g. nitrogen.

All vessels containing petrol must be correctly labelled and stored in a ventilated flammable liquid store.

It is advisable to avoid exposure to concentrated petrol fumes. Skin exposures should be treated immediately and contaminated clothing removed and left in fresh air to allow the contaminant to evaporate, ensuring this is carried out in a safe place.

## Occupational health

Workers who regularly handle petroleum spirit should receive pre-employment medical screening to ensure there is no history of skin disease that could be affected by exposure to the irritant and fat removing properties of petrol. It is suggested that workers who handle petroleum spirit that has a significant benzene content (or tetraethyl lead), should receive an annual occupational screening to exclude anaemia.

---

**First aid**

Inhalation

Petrol fumes inhaled can produce inebriation followed by unconsciousness. Remove the victim from the contaminated area into fresh air, the rescuers wearing CABA equipment. If the patient is unconscious, lie them on the ground and give artificial respiration using an oxygen-enriched resuscitator until breathing starts. Keep the victim warm and at rest. **Remember, if the victim has been working in an confined area with a petrol engine working, the victim may also be suffering from carbon monoxide poisoning. Refer the casualty to hospital without delay for specialist medical assessment.**

Ingestion

Acute poisoning can occur if petrol is ingested. The patient complains of oesophageal and stomach pain, and usually vomits. Ensure that the patient does not inhale the vomit and if conscious get them to rinse the mouth out with cold tap water. Do not induce vomiting, but give a glass of milk to buffer the effects of the petrol on the stomach lining, and delay the entry of stomach contents into the duodenum.

Eye and skin contact

Skin or eye contact with petrol must be treated immediately with a steady stream of cold tap water thoroughly washing the affected area for at least 15 minutes. **Any eye injury should be referred to hospital without delay for specialist medical treatment.** Unless the skin contact is trivial, the patient should be seen by a medically qualified person as soon as possible.

---

# Phosgene  COCl$_2$

## Carbonyl chloride

## Carbon oxychloride

**Exposure limits**
OES = 0.1 ppm (0.4 mg/m$^3$)

### Description
A colourless, non-flammable, highly toxic gas characterized by a smell of musty hay. It is normally supplied as a yellow liquefied gas in steel cylinders at 11 psi pressure.

### Properties
It is used in industrial synthetic processes often in large quantities.

MW = 98.9,   SG = 1.39,   MP = –128°C,   BP = 8°C,
VP = 760 mmHg (101.08 × 10$^3$ Pa),   VD = 3.4.

### Detection methods
Chemical gas detector tubes.
Infra-red gas analyser.

### Precautions when using the gas
**Phosgene is extremely poisonous when inhaled or if the liquid gas comes into contact with the skin.** Storage of gas cylinders must be in a dry, cool and well-ventilated area. If phosgene is used in synthetic laboratories inside buildings, a designated exhaust protective cabinet must be used, fitted with suitable gas absorbent filters in case of accidental release of the gas. **All staff handling or using phosgene must be fully trained in the safe handling of the gas and emergency procedures in case of accidental leakage. It is essential that respiratory protective equipment is available in the workplace.** This may take the form of full face canister type respirators, self-contained compressed air sets, or pressurized airline suits or hoods. **Staff working with phosgene should wear appropriate chemical resistant clothing, gauntlets and vinyl boots. On no account should staff work alone in a workplace using phosgene.**

### Occupational health
**The very low value for the OES for phosgene shows it is highly toxic.** In the 1914–1918 war, liquefied gas from cylinders was released to form vast clouds of the heavy phosgene vapour as a crude but effective form of chemical weapon; 80% of all military gas fatalities were due to phosgene. According to Donald Hunter (1975), **in low concentrations the gas causes little immediate irritation so that dangerous amounts can be inhaled before any toxic signs are seen. The gas is relatively**

insoluble in water so it is conveyed by the inspired air directly into the lungs. After some hours the victim may have a sudden onset of pulmonary oedema with circulatory collapse. Physical exertion by the victim will increase these problems.

---

**First aid**

Inhalation

Remove the casualty from the contaminated area into fresh air without delay, the rescuer(s) wearing breathing apparatus. Keep the victim at rest preferably lying on a stretcher and covered with a blanket to reduce shock and to minimize shivering. All physical exertions by the patient are to be avoided. If breathing is abnormal or distressed, give 100% oxygen or oxygen-enriched air by means of a resuscitator. **Because of the danger of the sudden onset of pulmonary oedema or respiratory collapse, a medical doctor should be called without delay and an ambulance arranged to take the casualty to hospital.** The ambulance paramedical staff should be informed of the precise nature and duration of the exposure.

Skin contact

Gas or liquid phosgene exposure must be treated immediately by removing the contaminated clothing and irrigating the surface of the skin with a stream of cold tap water for at least 15 minutes with the patient lying down. After washing, the casualty should be dried carefully and then placed on a stretcher and covered with warm blankets to reduce shock. The patient must be seen by a medical doctor if possible prior to transport to hospital.

Eye contact

**Gas or liquid phosgene in the eye is a serious matter.** Immediately irrigate the eye(s) with a slow stream of cold tap water for at least 15 minutes making sure that the water treatment reaches the orbit of the eye. Then place a dry eye dressing over the affected eye(s) **and immediately remove the casualty to hospital for expert medical specialist treatment.**

---

# Phosphine

# PH₃

## Hydrogen phosphide

**Exposure limits**
OES = 0.3 ppm (0.4 mg/m³)
STEL = 1.0 ppm (1.0 mg/m³)

## Description
**Colourless, flammable, highly toxic gas**, heavier than air, with a fishy odour.

## Properties
Phosphine is made from metallic phosphides and the phosphine gas stored in cylinders.

MW = 34, SG = 1.17, MP = −133°C, BP = −87.7°C,
IT = 40–65°C but may ignite at room temperature if dry.

## Detection methods
Chemical reaction tubes.
Wet silver nitrate paper.

## Precautions when using the gas
**Phosphine will ignite easily or react explosively in the presence of air.** Phosphine can be produced by allowing aluminium phosphate or calcium phosphide to become damp during storage. The most effective control measure is prevention of the build-up of the gas by efficient ventilation remembering that the gas is heavier than air.

**Personal respiratory protection is necessary for work with phosphine.** For low levels of the gas, a canister type gas mask would be suitable provided the breathable air contains adequate oxygen. Otherwise self-contained breathing apparatus is required.

Phosphine gas cylinders must be stored in well-ventilated, cool, dry isolated conditions.

(See HSE Environmental Hygiene Guidance Note EH20 (1979).)

## Occupational health
Because of the unpleasant smell of the gas, even at a concentration of 2 ppm, most exposures at work arise as a result of an accident or gas cylinder leak. The effects can be summarized as follows:

### Subacute effects
Inhalation of phosphine causes coughing, dyspnoea, and pulmonary oedema. Headaches with chest pains, followed by vomiting and diarrhoea may occur

depending on the amount of gas inhaled. Studies show that 7 ppm by volume of phosphine can be tolerated without ill effects for several hours.

## Acute exposure

Increasing exposure produces cyanosis, pulmonary oedema with ataxia and drowsiness. **If untreated, death may occur within 48 hours.** If the person survives there could be significant damage to liver, kidney, heart and brain; all these organs are particularly sensitive to phosphine.

---

**First aid**

**Remove the casualty from the contaminated area wearing compressed air breathing apparatus.** Lie the patient down on a stretcher or the ground and administer oxygen and artificial respiration if the patient's breathing is weak. Keep warm and at rest. **Arrange for the patient to be removed to hospital without delay**, informing the ambulance paramedical team of the type of poisoning present and the duration, if known, of the exposure. Early and effective first aid usually produces a full recovery in cases of phosphine intoxication.

No specific antidote to phosphine is known.

---

# Propane $C_3H_8$

**Exposure limits**
OES = 1000 ppm (1800 mg/m³)

## Description
A colourless, highly flammable gas forming very explosive gas/air mixtures.

## Properties
Heavier than air, highly flammable gas slightly soluble in water.

MW = 44.09,   VD = 1.6,   BP = –44.5°C,   EL = 2.2–9.5%,   IT = 466°C.

Propane is an easily compressible gas and is supplied in steel cylinders of various sizes containing the liquefied gas. Propane is used widely as a fuel gas source for industrial and domestic use.

## Detection methods
Flammable gas detector.

## Precautions when using the gas
**Propane in the liquefied form as liquefied petroleum gas is a major potential fire and explosion hazard in any factory, workplace or in the home.** Since the explosion limits in air are between 2.2 and 9.5%, and the gas is heavier than air and therefore not easily dispersed by ventilation, slight leakages from pipeline connections or cylinders can easily form an explosive mixture and give rise to explosions and fires. The gas should not be used in unventilated rooms or basements.

Note: 500 grams of the gas mixed with air has an explosive equivalent of 1 kilogram of TNT.

**Leakage of a standard size cylinder of liquid gas (25 kg) into a workshop or laboratory, allowed to mix with air and ignited would destroy the building and kill any occupants.**
Propane cylinders should be stored in the open air, protected from heat and sunlight, and the gas supply should be piped into the building, if possible, by means of fixed gas lines. Only small size gas cylinders, e.g. up to 6 kg, should be used inside buildings as a portable gas supply. **All compression joints and union couplings to cylinders and equipment must be soap tested for leaks daily, and every time the fittings are disturbed.**

## Occupational health

Concentrations of 10% propane in air (100000 ppm) do not cause irritation to the eyes or respiratory tract but may produce dizziness after inhalation for some minutes. Higher concentrations are not known to be toxic, although they may produce anaesthesia and will reduce the available oxygen content of respirable air.

---

**First aid**

Inhalation
The exposed person should be removed from the contaminated area into fresh air and allowed to recover. If the subject appears drowsy, supplementary oxygen should be administered until the breathing is normal and regular, and the subject appears to be well oxygenated as determined by normal pink coloured nailbeds or lips.

Skin contact
Liquid propane volatilizes readily from the surface of the skin but in significant amounts may produce a mild form of cold burn. If the spillage has saturated clothing with the liquid, the garments should be removed and the affected skin irrigated with a stream of cold tap water for at least 10 minutes. The skin should be examined and if there is severe irritation and redness with blisters, the casualty should have a dry burns dressing applied to the injury and be referred to the nearest Accident and Emergency department or appropriate Health Centre.

Eye contact
**Splashes of liquefied gas into the eye is a serious matter.** Immediately treat the casualty by irrigating the eye with a steady stream of cold tap water to prevent further tissue damage.

The irrigation should continue for at least 10 minutes. Cover the eye with a dry eye pad, do not place any medication into the eye, **and take the patient to hospital without delay**. Inform the medical staff at the hospital of the nature of the exposure and the substances involved.

---

# Radon

# $^{222}$Rn

## Exposure limits

The background exposure from radon is very small except in areas with known uranium or granite deposits. The atmospheric concentration will increase in processes involving mining or handling of material containing uranium. The OES for radon is $3 \times 10^{-8}\,\mu Ci/cm^3$ in air (ICRP 1959) and the maximum permitted annual intake during working hours is $730\,\mu Ci$ (IAEA, 1967).

## Description

Colourless, odourless radioactive gas produced as a transformation product in the decay of $^{226}$radium. This radionuclide has a life of 3.8 days before transforming into $^{218}$polonium.

## Properties

Radon is unstable and during its decay gives off high energy alpha particles, or helium atoms, with a potential of 5.48 MeV, and gamma rays. Both these radiations are harmful to humans and animals.

## Detection methods

Radon measurements are usually made by passing a large volume of air through a gas and particulate filter and then measuring the alpha radiation present using a scintillation counter. The concentration of radon daughter products can be determined by passing the air sample through a paper filter and then measuring both the alpha and gamma radioactivity present on the paper.

## Precautions when using the gas

Extreme caution is required to ensure that no significant radon gas or its daughter products are released into the air during any work process. All manipulations that might release radon or its products should be carried out in an exhaust protected enclosure.

The extracted air from this enclosure must be passed through special filters containing paper, glass fibre and active carbon to ensure partial retention of the radon but full retention of the more toxic disintegration products. Where this is not possible, high particle extraction filter masks (<5 µm) must be used.

## Occupational health

The decay pattern of radon shows five transformations to $^{210}$Pb each of which is of short duration, and at each stage high energy radiation is released in alpha, gamma and hard beta forms. Radon gas if inhaled is eliminated easily from the body via the expired air, about 90% in the first hour following exposure. Unfortunately the decay products of the radon, usually carried on dust particles, can easily be deposited in the respiratory tract emitting high radiation doses to the lung epithelium. These products, notably polonium, lead and bismuth can be deposited in selected organs such as the kidneys. **It is essential that all radon workers have a full employment** medical examination every six months which includes special attention to lung and kidney function and laboratory analysis of blood and urine. A measurement of the level of $^{210}$Pb should be made at each examination.

---

**First aid**

The exposed person should be removed from the source of contamination and any injuries such as fractures or bleeding treated using normal first aid techniques. Radon is easily absorbed from the lungs and being fat soluble widely distributed throughout the body. Radon is lost from the body via the lungs, some 90% eliminated in the first hour and the remainder in 7 hours. The main danger to the patient is the radon decay products ($^{210,218}$Po and $^{214}$Bi) which are present in the lungs' epithelia. These radionuclides are transported by the blood and selectively deposited in various organs, mainly the kidneys, and they present the major risk to the patient. These radon decay products are slowly eliminated from the body via the urine and faeces. **The patient must be transferred to hospital without delay so that remedial chelation treatment to remove the radon decay products can be carried out.**

---

# Stibine

# SbH₃

### Antimony hydride

**Exposure limits**
OES = 0.1 ppm (0.5 mg/m³)
STEL = 0.3 ppm (1.5 mg/m³)
Stibine is a highly toxic gas and should be handled with extreme caution.

## Description
Colourless, **highly toxic, flammable gas,** heavier than air, with a garlic like nauseating odour.

## Properties
Stibine may be produced accidentally if antimony compounds are treated with steam or if nascent hydrogen is passed over antimony or its compounds.

MW = 124.8,   SG = 4.36,   MP = −88°C,   BP = −17°C.

## Detection methods
Chemical detector tubes for arsine will react with stibine giving a weak grey-violet colour.

## Precautions when using the gas
**Inhalation of stibine gas is dangerous because the gas is extremely toxic and the minimum lethal dose is unknown.**

Work involving the use of stibine gas must be carried out in specially ventilated laboratories or workshops with the manipulations carried out in exhaust protective cabinets fitted with a filter that can remove stibine from the extracted air. **No work with stibine should be sanctioned until safe working arrangements are in place and have been tested.**

Stibine is usually supplied in small cylinders which should be stored in a lockable well-ventilated room. In the event of leakage from the stibine cylinder, the work area must be evacuated and staff must be kept away from the contaminated area, until the leakage has been stopped and the area ventilated.

**Trained staff using positive pressure breathing apparatus should attempt to shut the cylinder valve, and then move the defective cylinder into a filter cabinet or the open air, downwind from staff and buildings. The incident must be reported to the local fire brigade as a toxic chemical incident.**

## Occupational health
**Prevention of exposure to stibine is essential since accidents involving stibine gas can be severe or fatal.**

The toxicology of stibine is very similar to that of arsine.

## Acute exposure

Symptoms of stibine poisoning appear from 20 minutes to 3 hours after the initial exposure. The immediate symptoms are related to the effect on the central nervous system, e.g. giddiness, headache, tingling sensations in hands and feet, nausea and vomiting. Pain may also occur in the chest and abdomen. The urine of the victim, when produced, will be dark and blood stained since antimony is a powerful haemolytic agent. The patient's skin may appear coloured with a jaundiced hue. The toxicity of antimony is mainly due to its powerful action on enzymes in the body possessing SH groups. The antimony present will form very stable covalently bonded complexes with these enzymes.

Antimony (stibine) affects the red blood cells in two ways:

1. It combines with the SH containing tripeptide known as glutathione which is required to maintain the ion pumping mechanism of the red cell membrane. When this mechanism is disturbed, sodium ions from the plasma pass into the red cell causing osmotic swelling and finally rupture of the red cell membrane. This haemolysis, which can last up to 96 hours following antimony exposure, is responsible for the high plasma haemoglobin levels, leading to the production of the bile pigments, bilirubin and biliverdin, seen in jaundice and the blood-stained urine.
2. It produces a toxic antimony–haemoglobin complex by reacting with the SH groups on the haemoglobin protein molecule. This complex is non-dialysable by the kidneys and requires the technique of exchange blood transfusion for its removal.

Antimony toxicity affects many organs particularly the nervous system, heart, liver and the kidneys. The effect on the kidneys is very obvious, producing an antimony-induced anuria (failure to produce urine). This anuria may be due to either a direct action of the antimony on the tubular cells of the kidneys, or a massive release of the antimony–haemoglobin–haptoglobin complex which precipitates out into the lumen of the kidney tubules.

---

**First aid**

ANTIMONY POISONING IS AN ACUTE MEDICAL EMERGENCY.

**The patient must be removed to hospital by ambulance with minimum delay, preferably with medically qualified or paramedical staff in attendance.** The victim should be accompanied to hospital with a person who can explain the nature and extent of the poisoning to the hospital emergency staff. The hospital receiving this casualty should be alerted as to the nature of the exposure.

---

# Sulphur dioxide       $SO_2$
# Sulphur trioxide       $SO_3$

Sulphurous acid anhydride

Sulphuric acid anhydride

**Exposure limits for $SO_2$**
OES = 2 ppm (5mg/m$^3$)
STEL = 5 ppm (13 mg/m$^3$)
**Exposure limits for $SO_3$**
OES = 1mg/m$^3$
No STEL

## Description
Sulphur dioxide is a pungent, irritant, acidic and corrosive gas which is non-flammable.
Sulphur trioxide is a similar gas but with more toxic and corrosive properties. Both substances can exist in solid, liquid and gaseous forms.

## Properties
Sulphur dioxide readily dissolves in water to form sulphurous acid, which is slowly oxidized to sulphuric acid. The gas is also soluble in organic solvents. Sulphur dioxide is widely used as a bleaching agent in many industrial processes including the manufacture of paper from wood pulp. The gas is normally manufactured by heating iron sulphide in a stream of air.

MW = 64.06, SG = 1.43 (liquid), MP = −72.7°C, BP = −10°C, VD = 2.92, VP = 2538 mmHg (337.5 × 10$^3$ Pa) at 21°C.

Sulphur trioxide has similar properties to sulphur dioxide but is more chemically reactive. The trioxide is readily soluble in water, carbon disulphide and sulphuric acid producing a very powerful sulphating acid used in the synthetic chemical industry and known as Nordhausen sulphuric acid. These acids were used originally in Germany for the production of dyestuffs and explosives. Solid sulphur trioxide exists in three crystalline forms two of which are metastable and can explode.

MW = 80.06, SG = 1.92 (liquid), 1.97 (solid),

| for solid crystalline forms | alpha | beta | gamma |
|---|---|---|---|
| MP | 62.3°C | 32.5°C | 16.8°C |
| BP | 44.8°C | sublimes | 44.8°C |
| VD | 2.76 | 2.76 | 2.76 |
| VP = 100 mmHg (13.3 × 10$^3$ Pa) at | 10.5°C | 14.3°C | 62.1°C |

## Precautions when using the gas

Because many chemical processes use or produce sulphur dioxide or trioxide, contamination of the workplace is frequently found. To provide a safe and healthy work environment, great attention must be paid to the prevention of gas releases during manufacture and ensuring a good ventilation system so that the OES levels are not exceeded. Where possible the processes should be totally enclosed and any exhaust fumes or vapours passed through a water scrubber before being released to atmosphere. If this is not possible, then staff working in contaminated atmospheres must wear acid resistant clothing and full respiratory protection. This may necessitate the provision of positive pressure airsuits where sulphur trioxide enriched sulphuric acid is handled. To reduce the effects of these corrosive acid gases on the oral mucosa and the tooth enamel, EOHS recommends the use of drinking fountains and the rinsing out of the mouth with 10% sodium bicarbonate solution to neutralize any acid effects.

Safety showers and drenches are essential in all work areas handling these substances. Staff must not work alone in areas where these gases are used or produced.

## Occupational health

Sulphur dioxide is a very irritant gas due to its formation of sulphurous and sulphuric acids on contact with moist tissue and mucosa. The gas can enter the body via the respiratory tract and the mouth in the form of sulphurous acid, which rapidly enters the bloodstream. The increase in hydrogen ion concentration in the blood induces a metabolic acidosis stimulating compensation by the kidney in the formation of excess ammonium ions in the urine. The sulphate in sulphurous and sulphuric acid is excreted as sulphate ion in the urine. Exposure to high concentrations of sulphur dioxide and trioxide will produce generalized interference in the normal protein and carbohydrate metabolism, in some cases leading to damage to the haemopoietic system and the production of methaemoglobin. The oxidizing effects of sulphur dioxide can lead to destruction of vitamin C and the vitamin B complex leading, in long term exposures, to vitamin deficiencies.

Acute exposures are fortunately rare but give rise to intense irritation of the conjunctivae of the eyes and upper respiratory system, followed by dyspnoea and cyanosis, and unconciousness.

Death can result from reflex spasm of the larynx causing suffocation and circulatory failure.

Chronic exposures are more common and produce a persistent irritation and pain in the mucous membranes of the nose and throat. A long history of chronic exposure to high level of these gases can often be seen as chronic bronchitis with emphysema, leading to radiologically observed bilateral shadows. These acid gases also produce damage to the dental enamel and yellowing of the teeth.

Note: Sulphur trioxide behaves in a similar way to the dioxide but is much more chemically active. Sulphur trioxide liquid is extremely corrosive and will produce serious and painful burns to the skin and other tissues.

**First aid**

Inhalation

Sulphur dioxide and trioxide produce choking and coughing in exposed subjects. **Remove the victim from the contaminated area into fresh air, the rescuers wearing acid resistant clothing and suitable breathing apparatus. If breathing is difficult or stops, give artificial respiration preferably using a resuscitator with supplementary oxygen.** Keep the subject warm and at rest. Severe exposure may produce shock, rapid pulse, sweating and collapse. **Medical assistance is required as soon as possible.**

Skin contact

**If liquid dioxide or trioxide is seen, remove immediately by blotting the affected part with a dry cloth or paper. Then wash the surface of the skin with a slow stream of cold tap water from a flexible hosepipe or shower unit, removing the contaminated clothing as quickly as possible. Speed is essential to reduce the risk of severe acid burns.** In the case of sulphur trioxide, the clothing will disintegrate during this process. **Continue the water irrigation treatment for at least 15 minutes.** Gently dry the surface of the skin with absorbent paper. **If necessary cover the affected skin with a dry loose dressing and refer the casualty to hospital for a medical assessment of the injury.**

Eye contact

THIS IS A MEDICAL EMERGENCY.

**Immediately irrigate the eyes with a slow stream of cold tap water preferable from a flexible hosepipe ensuring that the eyes are thoroughly washed with water.** Encourage the patient to open and shut the eyes during irrigation to ensure all the acid gases or liquids have been removed. **If in doubt continue the irrigation process for a further 10 minutes. Cover the eyes with a loose dry dressing and refer the victim to the nearest Accident and Emergency hospital for expert opthalmological assessment.**

# Sulphuryl fluoride $SO_2F_2$

## Sulphuric oxyfluoride

**Exposure limits**
OES = 5 ppm (20 mg/m³)
STEL = 10 ppm (40 mg/m³)

### Description
**A very stable, colourless and toxic gas used as a fumigant.** It is supplied as liquefied gas in steel cylinders at 200 psi pressure.

### Properties
The gas is non-flammable, very stable and toxic. It is slightly soluble in water, but soluble in ethyl alcohol and carbon tetrachloride.

MW = 102, SG = 3.7, BP = −55.4°C, MP = −135.8°C.

### Detection methods
Infra-red gas analyser.

### Precautions when using the gas
**Because the gas is not detectable by smell, its presence in high concentrations in air cannot give early warning of a leakage.** Work with sulphuryl fluoride must be carried out in specially designated well-ventilated areas. All staff must be carefully trained in safety procedures in the event of a gas leakage, to protect the workers and persons who may be in the vicinity of the workplace. The provision of chemically resistant clothing, eye, hand and foot protection is required. Because of the toxicity of the gas, self-contained breathing apparatus should be readily available for use in case of an emergency. The provision of eye wash stations and showers are required for this work.

### Occupational health
Animal experiments involving exposure to this gas indicate that the OES quoted for humans should not be exceeded. Accidents in chemical works where workers have been exposed to the gas for about 4 hours show victims with nausea with vomiting, abdominal pain and severe skin irritation in the contaminated workplace. Following rescue and recovery, casualties show reddening of the nasal and pharyngeal mucosae and parathesia in the leg. The gas can produce pulmonary oedema if prolonged exposure to it occurs.

**First aid**
There is no specific treatment for sulphuryl fluoride exposure.

Inhalation
Remove the victim from the contaminated area into fresh air immediately, the rescuer(s) wearing self-contained compressed air breathing apparatus. If the casualty is breathing normally, place face downwards with the head lower than the lungs, to reduce the possibility of pulmonary oedema. **If breathing is abnormal, give 100% oxygen, or use an oxygen-enriched resuscitator until normal breathing is restored. The patient should be seen by a medically qualified person immediately after first aid treatment to assess the severity of the exposure. Delayed respiratory and central nervous system depression can occur.** If this is not possible, arrange for the casualty to be removed to hospital by ambulance. Give full details to the ambulance staff of the nature and duration of the exposure.

Eye or skin contact
Following removal of the victim from the contaminated area, irrigate the eyes and/or the skin with a stream of cold tap water for at least 15 minutes. Remove any contaminated clothing while the water drenching is being carried out. **Following first aid, the casualty should be taken without delay to hospital where further specialist treatment can be given.**

# Unsaturated aliphatic hydrocarbons

Ethylene $CH_2=CH_2$

Propylene $CH_3CH=CH_2$

Butylene $CH_3CH_2CH=CH_2$

Butadiene $CH_2=CHCH=CH_2$

## Description

These olifine gases are a series of chemically reactive unsaturated hydrocarbon compounds highly flammable and explosive when mixed with air or oxygen.

## Properties

Ethylene, a colourless gas with a sweetish odour

MW = 28, D = 0.61, MP = −169.4°C, BP = −104°C, VD = 0.98, EL = 3.1–36%, IT = 450°C.

Propylene, a colourless gas

MW = 42.1, D = 0.58, MP = −185°C, BP = −47.8°C, VD = 1,5, VP = 7600 mmHg (1011 × $10^3$ Pa) at 19.8°C, FP = −108°C, EL = 2.0–11.1%, IT = 497.

Butylene, a colourless gas with a slight odour

MW = 56.1, D = 0.66, MP = −185.3°C, BP = −6.3°C, VD = 1.9, VP = 3480 mmHg (462.8 × $10^3$ Pa), FP = 108°C, EL = 1.6–9.3%, IT = 384°C.

**Butadiene, a colourless, flammable gas with a pungent aromatic odour. It is highly reactive with a tendency to polymerize**

MW = 54.1,  SG = 0.65 at $-6.0$°C,  MP = $-108.9$°C,  BP = $-4.4$°C,  VD = 1.9,
VP = 760 mmHg ($101.08 \times 10^3$ Pa) at $-4.5$°C,  FP = $<7$°C,  IT = 428.9°C,
EL = 2.0–11.5%.

## Detection methods
Infra-red gas analyser.
Chemical reaction tubes.

## Precautions when using the gas
**The greatest danger from hydrocarbon gases is fire.** The plant area or room where the gases are used should have good ventilation. In industry where large quantities of these gases may be stored, a modern gas leakage alarm system should be installed which will automatically warn of any significant release of flammable gas. Special fire fighting equipment, e.g. water sprinkler system, will be required to contain and extinguish any fire. The type of storage facilities and the fire control measures needed can be calculated from the total flammable load in the storage area(s) as part of the risk assessment of the chemical process. To minimize the danger from fire the atmospheric concentration should not exceed one-fifth of the lower explosive limit, i.e. <0.4%.

## Occupational health
Occupational studies of workers exposed to butadiene show that concentrations of up to 8000 ppm for 8 hours have been tolerated with no serious ill effects. Slight irritation of the eye, nose and throat was reported at this concentration.

---

### First aid

Inhalation
Workers exposed to high concentrations of these gases may show symptoms of general anaesthesia or drowsiness. They should be removed from the contaminated area into fresh air. Make sure that the subject's airway is clear. If the patient's breathing is poor or has stopped, give artificial respiration (mouth to mouth) or an oxygen-supplemented resuscitator if available.

Skin contact
**Liquefied hydrocarbon gas can produce serious cold burns.** Treat the area affected by immersion in warm, not hot, tap water for at least 5 minutes. Carefully dry the area and apply a dry dressing and treat as for a thermal burn. Seek medical advice.

Eye contact
**EYE CONTACT WITH LIQUEFIED HYDROCARBON GAS IS A MEDICAL EMERGENCY.** Immediately irrigate the eyes with a stream of cold tap water for at least 15 minutes. Place a dry eye pad dressing over the eye(s) **and refer the casualty to hospital without delay for specialist medical treatment.**

---

# Vanadium and compounds V

**Exposure limits**
OES (TWA) = 0.5 mg/m$^3$ (ceiling) for total inhalable dust
OES (TWA) = 0.05 mg/m$^3$ for fume and respirable dust

## Description
Light grey lustrous powder, stable, does not oxidize in air.

## Properties
Vanadium has chemical similarities to nitrogen and arsenic. It forms compounds and oxides which have valencies of $2^+$, $3^+$, $4^+$, and $5^+$. The two most important vanadium compounds used in industry are vanadium pentoxide and ammonium metavanadate.

### Vanadium pentoxide (V$_2$O$_5$)
MW = 181.88,   SG = 3.36,   MP = 690°C,   BP = 1750°C
red-yellow rhombic crystals which are slightly soluble in water.

### Ammonium metavanadate (NH$_4$VO$_3$)
MW = 116.98,   SG = 2.33,   MP = 200°C
white crystals slightly soluble in water and dilute ammonia.

## Industrial uses
Large amounts of vanadium are used to produce ferrovanadium, a tough steel used for the production of high speed tools and high tensile steels. The vanadium added removes oxygen and nitrogen from the molten steel and improves the oxidation resistance of the resultant product. Vanadium pentoxide is important as an industrial catalyst in the production of sulphuric acid, metallurgy and photographic developing reagents.

## Detection methods
Colorimetric analysis using reagents which form coloured complexes with vanadium, e.g. using 8 hydroxyquinoline and extracting the vanadium complex from the interfering metals.
Phosphotungstovanadic acid method.
Gas chromatography using acetylacetone to form a complex which is separated from other metal acetylacetonates.

Atomic absorption spectroscopy either with a flue-rich reducing flame or using flameless carbon furnace with chelation-extraction pretreatment.

## Precautions when using the substance

**The low values for the Occupational Exposure Standards for the dust and fumes from vanadium show the importance of airborne and respirable dust suppression in manufacturing and industrial processes.** Because of the presence of significant amounts of vanadium in crude oil, exposure to vanadium can occur during the cleaning of oil-fired power station chimneys and ships exhaust flues since oil burning produces soot which contains appreciable amounts of vanadium. Workers carrying out these processes must wear protective clothing and full face respiratory protection.

## Occupational health

**Inhalation of vanadium containing dust will produce serious toxic symptoms even after brief exposures, e.g. 1 hour.** The early effects following a few days of exposure are seen as lacrimation, burning sensations in the eyes, throat, and mucous membranes leading to a productive cough, chest pain and bronchitis.

Severe acute exposure can give rise to a fatal pneumonia; less severe exposure will produce early effects which normally resolve after 2 weeks. Vanadium compounds often produce a green discoloration of the tongue which can be used as an indication of exposure.

Inhaled vanadium dust is taken up by the body and is excreted in the urine and faeces. It is not a cumulative poison and recovery from its effects usually takes place in one or two days. Workers who are to be employed in the handling or processing of vanadium products must have a pre-employment medical examination to exclude persons suffering from chronic respiratory illness. They should also undergo regular periodic medical examinations to exclude chronic exposure to vanadium.

**First aid**

Inhalation

The breathing of vanadium dust or fumes produces irritation to mucous membranes leading to cough and chest pains. The casualty should be removed as soon as possible from the contaminated area into fresh air, and allowed to rest with the head and thorax raised to reduce the coughing and improve breathing. In acute exposure cases it may be necessary to give oxygen-enriched air via a resuscitator. The patient should be kept comfortable and warm but not hot. All cases of inhalation exposure should be referred to hospital for expert medical examination and if necessary further treatment.

Ingestion

If vanadium pentoxide or vanadium salts have been ingested it is important to try and prevent the material passing into the bowel from the stomach. If the patient is conscious, their mouth must be washed to remove any solid material, and then a litre of water or milk given to drink followed by an emetic to encourage vomiting. (Note: Vanadium salts are not readily absorbed from the gut and may themselves produce vomiting if ingested in sufficient quantity.) The patient should then be referred to hospital for examination by a medical specialist.

Skin contact

Vanadium metal powder or salts should be removed from the surface of the skin by thorough irrigation of the affected area with a stream of cold tap water for at least 10 minutes. Skin rashes may appear on exposed areas such as hands and the face, and these should be treated by referring the patient to a medical practitioner.

Eye contact

Any vanadium metal, dust or salts affecting the eyes must be removed carefully by irrigating the eyes with a slow stream of cold tap water for at least 15 minutes. The eyes should be covered with a dry dressing and the patient referred to the nearest hospital for specialist treatment.

Note for medical staff: Some hospitals use irrigation with buffered phosphate solution followed by instillation of liquid paraffin drops as the follow-up treatment to initial first aid.

# Vinyl chloride ## CH$_2$=CHCl

**Exposure limits**

OES = 7 ppm (long-term exposure limit –
8 hour time weighted average concentration)

**IMPORTANT NOTE**

In the UK, vinyl chloride is subject to a one year reference period for work in enclosed vinyl polymerization plants. For all staff in these areas, continuous or permanent sequential air sampling methods must be used. From this data, the mean annual maximum exposure limit must not exceed 3 ppm.

## Chloroethylene

## Chloroethene

## Description
A colourless, highly flammable gas, having a sweetish smell at high concentrations.

It is supplied in steel cylinders as a liquefied compressed gas at its vapour pressure.

## Properties
Vinyl chloride monomer is the basic material for the production of the plastic, polyvinylchloride. This process involves the synthesis of some 10 million tonnes per year. **The liquid and gas are both flammable and form explosive mixtures with air. Fires involving vinyl chloride give rise to the toxic products, hydrochloric acid and carbon monoxide.** The gas is slightly soluble in water, soluble in ethyl alcohol, but very soluble in ether and carbon tetrachloride.

MW = 62.5,  SG = 0.91,  MP = –159.7°C,  BP = –13.9°C,  VD = 2.2,
VP = 2604 mmHg (345.8 × 10³ Pa),  FP = –78°C,  EL = 4 to 22%,  IT = 472°C.

## Detection methods
Chemical gas detector tubes.
Infra-red gas analyser.

## Precautions when using the gas
**Because vinyl chloride is both flammable and toxic at low concentrations, great care must be exercised in its use particularly in industrial locations.** The workplace must be carefully controlled, a full risk assessment must be made as to the degree of risk associated with each process involving the monomer. **One of its greatest dangers is the fact that the threshold of smell for the substance is in the range 2000–5000 ppm so that dangerous concentrations of the order of 100 ppm are not perceived.** Air extraction from the work areas must be monitored continuously to ensure the exposure concentration limit is not exceeded. In all work areas, chemical

resistant clothing with hand, foot and respiratory protection are required. In high risk areas, the use of airline pressurized fully protective suits may be needed.

## Occupational health

**The effects of long-term industrial exposure to vinyl chloride are well known and referred to as 'vinyl chloride disease'.** This condition affects the nervous system by the appearance of neurotoxic symptoms, peripheral microcirculatory changes similar to those of Raynaud's phenomenon, skin and skeletal changes, fibroses of the liver and spleen, and a pronounced carcinogenic effect leading to angiosarcoma, mesenchymal or endothelial carcinomas. According to the ILO Geneva, some 90 cases of angiomatous mesenchymoma have been observed. They show a delayed onset or incubation period of 20 years reminiscent of the course of the mesothelioma of the pleura seen in cases of blue asbestos exposure. **Because of the very serious nature of exposure to vinyl chloride, and the vast use of this product in industry, regular surveillance protocols should be set up for workers who may be exposed.** These involve tests of lung and circulatory functions together with liver function tests measuring serum transaminase activity, and screening for the level of metabolites in the urine derived from vinyl chloride such as thiodiglycolic acid.

---

**First aid**

Inhalation

The casualty should be removed without delay from the contaminated area, the rescuer(s) wearing protective clothing and personal protective breathing apparatus. Take the victim into fresh air and if possible place on a stretcher, keeping them warm with a blanket and quiet. If breathing is affected administer oxygen-enriched air from a rescuscitator. **Summon medically qualified help as soon as possible and call an ambulance to remove the casualty to hospital for specialist medical treatment. Ensure that the ambulance crew know the nature and type of exposure caused to the patient.**

Skin contact

Immediately drench the affected area with a stream of cold tap water while removing the contaminated clothing and shoes. Continue the irrigation treatment for at least 10 minutes, followed by further washing of the affected areas with soap and water. **The patient should be taken by ambulance to hospital for medical treatment.**

Eye contact

Irrigate the eyes with a steady stream of cold tap water for at least 15 minutes, making sure that the whole of the surface and tissues surrounding the eye(s) are fully washed. Cover the eye(s) with a dry eye pad and **refer the casualty to the nearest hospital for further specialist medical treatment**.

---

# Zinc and compounds     Zn

## Exposure limits

Zinc oxide (fumes)     OES (TWA) = 5 mg/m$^3$
                       STEL = 10 mg/m$^3$
Zinc chloride (fumes)  OES (TWA) = 1 mg/m$^3$
                       STEL = 2 mg/m$^3$
Zinc chromate (as Cr)  OES (TWA) = 0.05 mg/m$^3$
(this substance has been defined as a carcinogen for the purposes of the
COSHH Regulations).
Zinc phosphide   No exposure limits quoted. The substance is known to
be a systemic poison and is toxic to humans independent of the route of
entry into the body. Zinc chloride and zinc phosphide are the two most
toxic of the zinc salts.

## Description

Zinc exists as the free metal and many soluble and insoluble salts which have very important industrial uses. Some 5 million tons of zinc metal is produced from zinc ores every year by roasting and then refining to a pure form by electrolysis. In this form zinc is used for the preparation of alloys.

## Properties

Zinc metal (Zn) is a soft, silvery white metal, easily soluble in mineral acids to produce salts, such as zinc chloride. If the metal is heated above 500°C, the liquid zinc readily forms volatile fumes of zinc oxide.
AW = 65.4,   SG = 7.14,   MP = 419.4°C,   BP = 907°C.

**Fire Safety Note: Powdered zinc, if stored in quantity, can give rise to explosion or spontaneous combustion if water is present, and constitutes a potential fire hazard. The material must be stored in cool, dry, well-ventilated areas, away from oxidizing agents. Extinguishing fires involving zinc metal require a special type of dry powder containing fire extinguisher, which must be made available near to the storage site.**

Zinc oxide (ZnO) is an odourless white powder, soluble in strong acids and alkalis, but insoluble in water and alcohol. .
MW = 81.4,   SG = 5.6,   MP = 1975°C.

Zinc chloride (ZnCl$_2$) exists as white granular deliquescent crystals, formed from the action of hydrochloric acid on zinc or zinc oxide, and is soluble in water, ethyl alcohol and ether.
MW = 136.3,   SG = 2.91,   MP = 283°C,   BP = 732°C.

**Zinc chromate (ZnCrO$_4$)**, an insoluble yellow powder, soluble in acids but insoluble in water. It is formed by the action of chromic acid on zinc oxide or hydroxide.
MW = 181.4,   SG = 3.4.

**Zinc cyanide (Zn(CN)$_2$)**, a white powder, insoluble in water and alcohol, but soluble in dilute mineral acids with the evolution of HCN gas.
MW = 117.4,   SG = 1.85,   MP = 800°C (with decomposition).

**Zinc sulphate (ZnSO$_4$.7H$_2$O)**, colourless crystals, soluble in water and glycerol, produced by the action of sulphuric acid on zinc or zinc oxide.
MW = 297.4,   SG = 1.96,   MP = 100°C,   BP = 280°C.

**Zinc phosphide (Zn$_3$P$_2$)**, dark grey cubical crystals, soluble in benzene and carbon disulphide but which react violently with water, acids and oxidizing agents producing phosphine gas.
MW = 258.1,   SG = 4.55,   MP = 420°C,   BP = 1100°C.

## Industrial uses
**Zinc metal** is used in the galvanizing of steel to prevent corrosion and in the cathodic protection of ships hulls and steel tanks. Pure zinc metal is used extensively in the preparation of alloys.
**Zinc oxide** is used widely in the preparation of paints, as a filler material for plastics, and in the cosmetic and pharmaceutical industries.
**Zinc chromate** is used in the preparation of pigments, varnishes and lacquers.
**Zinc cyanide** is used as a plating agent and, because of its toxicity, as a pesticide.
**Zinc sulphate** is used as a preservative, as a bleaching agent for paper, and as a fungicide.
**Zinc phosphide** is very toxic with a smell which is attractive to rodents. It is widely used in rat poison preparations.

## Detection methods
There are many good methods for the detection and estimation of zinc. Zinc present in air samples or as fumes can be determined by extraction into solvents and determination by one of the analytical methods described:

Colorimetric analysis – diphenylcarbazone and di-beta-naphthalthiocarbazone are two reagents which readily react with zinc to form a coloured complex which can be determined spectrophotometrically.
Anodic stripping voltammetry has been used to separate and quantitatively measure zinc in the presence of zinc and other heavy metals in urine samples.
Atomic absorption flame spectroscopy has been widely used to measure the zinc concentrations present in liquids and biological fluids. More recently, flameless atomic absorption using chelation pretreatment of the samples has been used to measure very low concentrations of zinc.

## Precautions when using zinc and zinc compounds
Molten zinc at >500°C must be handled with care in industrial processes. Because the boiling point of zinc metal is 900°C, the molten zinc galvanizing process does not produce metal fumes that could give rise to metal fume fever. If galvanized metal sheets are subjected to an acid pickling process, care must be taken to avoid possible liberation of arsine gas in the process due to the presence of arsenic impurities present in the zinc.

Work with hot zinc metal or zinc chloride requires the use of protective screens, clothing, safety visors and goggles. Welding processes involving brass, because of the high temperatures used, may give rise to metal fume fever. The safety equipment described together with adequate local exhaust ventilation are essential to avoid this hazard.

## Occupational health

Although zinc is an essential trace metal required for normal human nutrition, the element can easily enter the body by inhalation, ingestion or skin absorption. As already mentioned, heating zinc metal or its oxide to high temperatures can lead to the production of metallic fumes which may produce metal fume fever. Four zinc salts are particularly dangerous to handle: zinc chloride, zinc chromate, zinc cyanide and zinc phosphide. Zinc chloride fumes are particularly irritant to the eyes and mucous membranes. Accidents have been reported when smoke generators using zinc chloride have been used. In the fatal cases, death was due to pulmonary oedema and bronchopneumonia. Zinc chromate used in primer paints in the car manufacturing industry has led to nasal ulceration and dermatitis. The caustic action of zinc chloride when used for wood treatment or as solder flux has led to ulceration of the fingers, hands and forearms of exposed workers.

---

### First aid

#### Inhalation
Inhalation of zinc dust or fumes can give rise to irritation of the eyes, nose and mucous membranes. In severe cases, where the worker has been exposed to fumes from boiling zinc or welding fumes, metal fume fever symptoms may be seen.

**It is important to remove the victim from the contaminated area as soon as possible and to then assess the degree of severity of the exposure.**

In cases of mild exposure, the patient will appear flushed and with obvious irritation to the mucous membranes in the nose and throat. These patients should be placed in a comfortable position, at rest, and reassured. They should make a full recovery in a few hours. If in doubt, the patient should be seen by a medically qualified person.

In more severe cases, patients may show signs of metal fume fever. Symptoms include shivering, irregular fever, profuse sweating, nausea, thirst, severe headache, and a feeling of exhaustion. In these cases, the patient must be placed at rest and kept warm but not hot to reduce shock, and either seen as soon as possible by a medically qualified doctor or taken to hospital for further assessment. Most cases make a complete recovery in 24 hours from the time of exposure.

#### Ingestion
Zinc salts are astringent and corrosive and in higher concentrations, if ingested, produce vomiting. The taste threshold for most zinc salts is about 15 ppm so toxic doses are unlikely to be ingested by accident. Occasionally zinc salts enter food during its preparation in worn galvanized vessels. The victim of this type of ingestion suffers from stomach cramps, vomiting, diarrhoea, usually occurring soon after the ingestion of the zinc salt.

The victim should be taken to hospital for a gastric lavage without delay after drinking a glass of milk. This buffers the acidic action of the zinc salt, and delays the passage of the contents of the stomach into the duodenum.

### Skin contact

Molten zinc in contact with skin produces severe burns and the metal may adhere to the skin making it difficult to remove. Immediately drench the affected area with a stream of cold tap water for at least 10 minutes in order to reduce the damage to the underlying tissues caused by the heat from the metal. Treat the victim for the effects of the severe burn by covering the injured part with a clean towel or clingfilm to reduce infection, and keep warm and at rest to reduce shock. **Arrange for the injured person to be taken to hospital without delay to receive specialist treatment. On no account put any antiseptic ointment or burn cream or try to remove metal from the damaged skin area.**

With zinc salts, the skin should be treated with large volumes of cold tap water for at least 10 minutes ensuring that all the salts have been removed. Zinc chloride can produce caustic burns to the skin seen as severe reddening of the skin. This type of injury is best treated by the Accident and Emergency department of the local hospital, and the patient should be taken there by car.

**Zinc cyanide exposure should be treated as a serious matter.** The victim should be treated with large volumes of cold tap water to wash off as much cyanide as possible and prevent absorption through the skin. The person involved should be kept at rest and warm **and taken to hospital without delay for further medical specialist assessment and treatment.**

### Eye contact

**Molten splashes of zinc into the eye is a medical emergency and usually means coagulation of the eye proteins and loss of sight.** The victim should be treated with large volumes of cold water to reduce the heating effect of the metal splash, **and then referred immediately to hospital for specialist treatment.**

Zinc metal dust, salts or zinc fumes in the eye should be treated by irrigating the eye and surrounding tissues with a stream of cold tap water for at least 10 minutes. This procedure often removes the foreign particles from the cornea and orbit of the eye. The patient should then be taken to the nearest Accident and Emergency department for further treatment and assessment of any injury.

# Index